The Future

D1486116

to Edward, Helen, Richard, Deborah

The Future

An Essay on God, Temporality and Truth

J. R. Lucas
Fellow of Merton College, Oxford

No philosopher of the first or second rank has defended fatalism
or been at great pains to attack it.

Gilbert Ryle

Basil Blackwell

Copyright © J. R. Lucas 1989

First published 1989

Basil Blackwell Ltd
108 Cowley Road, Oxford, OX4 1JF, UK

Basil Blackwell Inc.
3 Cambridge Center
Cambridge, Massachusetts 02142, USA

British Library Cataloguing in Publication Data

Lucas, J. R. (John Randolph), *1929–*
   The future: an essay on God, temporality
   and truth.
   1. Time. Philosophical perspectives
   I. Title
   115

   ISBN 0–631–16659–9

Library of Congress Cataloging in Publication Data

Lucas, J. R. (John Randolph), *1929–*
   The future: an essay on God, temporality, and truth/J. R. Lucas.
      p.   cm.
   Includes index.
   ISBN 0–631–16659–9
   1. Time.   2. God.   3. Truth.   I. Title.
   BD638.L83   1969
   115—dc 19   88–27541
                                                          CIP

Typeset in 11 on 13 pt Sabon
by Opus, Oxford
Printed in Great Britain by Billing & Sons Ltd, Worcester

Contents

Preface

I have often tried to think straight about time, but never successfully. As an undergraduate I acquired from Gilbert Ryle the courage to believe what I already knew, and a certain robustness of approach towards the fallacies of fatalism. Later it was my great good fortune to attend the lectures of Arthur Prior, and begin to see how logic could be extended to deal with time and tense. The work of Saul Kripke was a revelation, showing how different modal logics could be understood in terms of the accessibility relation, and I owe a debt of gratitude to G. E. Hughes for making me aware of it. I tried out my ideas in papers at Oxford, Cambridge, Aberdeen, St Andrews, and elsewhere, and am grateful to Ian Crombie and others who responded to them over the years, and to J. Almog, then a research student at Merton, who at one juncture provided me with crucial bibliographical references. I have learned much from Tony Kenny, often indeed in being forced to think hard what my reasons were for disagreeing with him, but much all the same. Christopher Kirwan convinced me of the significance of Cambridge change in the course of a class we gave on St Augustine. In 1981 I was invited by the University of Dundee to give the Margaret Harris Lectures on Religion, and took the opportunity to develop the theological implications of what I was thinking under the title 'God, Temporality and Truth'. In 1987 I returned to the same theme in a lecture at Wheaton College, Illinois, and in the Harry Jellema Lectures at Calvin College, Grand Rapids, Michigan, where I was exposed to pertinent questioning from Nicholas Wolterstorff whose theological thinking had been tending in the same direction as mine over many years. Over many years, too, Storrs McCall had been thinking about branching time, and in a bar in Montreal gave me a key argument for chapter 10.

I do not think I have got time straight yet; in particular, I suspect that there are further arguments about the modal derivation of time I have not been able to catch hold of, and which could lead us to a better appreciation of both the logic and the ontological status of time. The theological implications have only been sketched, and need much more working out, and, in particular, the similarities with, and points of divergence from, the views of the Process Theologians. But time is short, and the best is the enemy of the possible. I ought at this stage to go public, sharing any good ideas I have with a wider audience, and exposing my bad ones to further criticism.

Anyone writing a work on logic has to make choices between rigour and intelligibility. I have sought the latter. Much rebarbative symbolism remains, but I have tried to gloss it throughout in ordinary English, and to conduct the argument in informal as well as formal terms. The symbolism is intended to make evident distinctions otherwise difficult to draw, and to sharpen the subsequent argument, but I have drawn back from pursuing many interesting questions formal logicians are interested in. They will find several avenues unexplored, and no doubt many holes in the argument as presented. But anyone acute enough to see the holes will be able also to see how they may be blocked – if they can.

In spite of the acknowledgements I have made, there remains a difficulty about acknowledgements and references generally. The topics I have been treating have been treated by many other philosophers down the ages, by many in our own time. I have not discussed them all. I have not read them all. If I had even attempted to read all that was relevant, I should have had no time to think things through myself. Where I am conscious of having appropriated another man's thought, I have acknowledged it, and where I reckon that it would be helpful to the reader to pursue a point, or where fairness requires it, I have given references. But often I have passed over the contributions or objections of others in silence. I can only apologize to them, and say that it is more likely to be due to my ignorance of what they have had to say than to a low regard for its value.

There is always a problem of dating in a work about The Future: for the future is constantly becoming present and then past. At each successive revision of the book, I have had to change the dates, adding on another seven years. I have therefore fixed on a date for tomorrow's sea battle, which should remain in the future for most of my remaining life. It is February 29th, 2000. It is a very special day. It is a leap-year's day: more than that, it is a leap-year's day in a year divisible by 100; most centuries – 1700, 1800, 1900 – are not leap years, even though they are divisible by 4; but if they are divisible by 400, they are. So February 29th, 2000, is only the fifth such leap-year's day we could have had in our era, if we had always used a modern calendar, and at any rate the first millennial one we have ever had. It is the most unique day that our essentially non-particularizing dating scheme can offer in the lifetime of readers alive when I wrote the book: whether I shall be alive to see that day, and more importantly, whether it will in fact be graced by a sea battle, only time will tell. So tomorrow is February 29th, 2000, from which it follows that today is February 28th, 2000; often I shall use the subscripts t and u to indicate today and tomorrow respectively.

Besides choosing a time, it is also necessary to indicate a place of utterance. I have chosen Cambridge. In part it is a precaution against the parochialism of philosophers, who take most of their examples from their study or the Quad, and believe that the A40 is universally the best road to London. In part it is a tribute to a university which since Barrow and Newton has done much to deepen our understanding of time. Cambridge philosophers and scientists have often held wrong views about time, and I, in a manner of speaking, could be represented as maintaining that time is much more real than many of them would allow. But since I am following contemporary usage, and meaning by 'Cambridge change' a change that is not a real change, let me redress the balance by siting myself in a place that is indubitably real.

J. R. L.

1

Time and Reality

The future is a touchstone for our attitudes to time and reality, to causality and freedom, to responsibility and creativity. If we believe that the future exists in just the same way as the present and past do, we have a simple and economical view of reality, but one in which many of the characteristic features of time are lacking, in which there is nothing to pick out the present 'now', or to give a sense of becoming, in which freedom is an illusion, everything will be as it will be, and there is nothing new under the sun. If, on the other hand, we believe that that the past and present exist but not the future, except as some set of tenuous possibilities, then we begin to understand why the past is unalterable and the future open, and have a view of reality that accounts for the peculiar status of the present and our sense of time as becoming. It allows for freedom and responsibility and creativity, and since they cannot be undone or conjured out of existence, it acknowledges the everlasting significance of our deeds.

These are not the only possibilities. Some have held that only the present is real, and neither the past nor the future exists. Only the present is actual, as French usage suggests. The past no longer exists and the future does not yet exist, and both are real only in so far as they are relevant to real experience or real decision here and now. Aristotle knew of that position,[1] and St Augustine argues for it with great force.[2] In modern times

---

[1] Aristotle, *Physics*, IV, 218a.
[2] St Augustine, *Confessions*, XI, xvii-xviii.

Lukasiewicz and Dummett have both argued for the unreality of the past as well as of the future.[3] It is an attractive doctrine, based on the insight that the present is essential for the reality of time generally, and it carries an important message for an age in which men are eager to mortgage the present to the future, as well as often being also the prisoners of the past. We need to be told to gather rosebuds while we may, because we are too ready to postpone enjoying life until we have done A-levels, got through Prelims, taken Finals, got a degree, got a job, got promotion, finished the job, cleared our desk, paid off the mortgage, retired, died . . . *Carpe diem*. If it is never jam today, it is just simply never jam: only on a today can we enjoy ourselves, and if we will not enjoy today, we shall not enjoy at all. Yet this, although true, is not the whole truth. Although every day must at some time or other be a today, the word 'today' has the meaning it does have only because we know that there are other days which are not today today; we may take no thought of the morrow, but we cannot make out that it will not come, bringing its sufficiency of evil with it, and we should not be doing justice to a man's individuality as an agent if we reckoned that all his yesterdays were of no account.

The latter consideration tells also against the doctrine that the past does not exist, but the future does. It has few defenders among metaphysicians, but many practitioners among utilitarian moralists. It is fashionable to be future-oriented, and to regard everything past as water under the bridge and of no continuing significance. Utilitarians are systematic consequentialists, evaluating actions entirely with regard to their future consequences, but have great difficulty in saying what an action

---

[3] Jan Lukasiewicz, *Collected Papers*, ed. L. Borkowski, Amsterdam, 1970, pp. 127–8; reprinted in Storrs McCall, *Polish Logic*, Oxford, 1967, pp. 38–9; quoted by A. N. Prior, *Past, Present, and Future*, Oxford, 1967, pp. 29–9; M. A. E. Dummett, 'The Reality of the Past', *Proceedings of the Aristotelian Society*, 69, 1968–9, pp. 239–58; reprinted in M. A. E. Dummett, *Truth and Other Enigmas*, London, 1979, ch. 21, pp. 358–74; and 'Bringing about the Past', *Philosophical Review*, 73, 1964, pp. 338–59, reprinted, in Dummett, *Truth and Other Enigmas*, ch. 19, pp. 333–50.

is or what its consequences will be, except in the context of
what has previously been done. Moreover, since the future will
be past, it is difficult to take it that seriously, if it is doomed to
go out of existence. Coherent consequentialism is hard to
sustain, if consequences cease to count once they have occurred.

Nevertheless, the present is peculiarly significant for the
existence of time. Unless we can connect a time sequence with
the present, it ceases to be real: if our only answer to the
question 'when?' is the studiously vague 'once upon a time', we
are not talking about real facts at all, but only fiction. Time is in
this respect different from space. It is perfectly intelligible to
envisage a space in which we are not located: we can
understand a geometer when he draws on the blackboard, and
there is no question of any point's representing 'here', but not a
historian who puts up a date chart on which there could be no
date that is 'now'. The fact that we cannot conceive of a time
totally divorced from our own temporal existence is the
converse of time's being a necessary condition of consciousness
and experience, whereas space is not.[4] A sense of the peculiar
importance of the present underlies McTaggart's attempt to
prove the unreality of time, though he formulated his argument
in terms of past and future as well as present; McTaggart tried
to show that no time could consistently be described as the
present time, and though, as I maintain, his attempt fails, the
direction of his attack shows the crucial significance of tense,
and thus of the present, for the ontological status of time.[5]

But once again, the reality of the present, essential though it is
to the reality of time, does not prove that the rest of time is
unreal. Today is a day, and there are other days that were each
today in its day. The present may be pre-eminent, but cannot
stand alone; and if it is real, reality must be accorded to some
other time too. Moreover, if the past and the future were

[4] See more fully, J. R. Lucas, *A Treatise on Time and Space*, London, 1973,
§§2,7,52 pp. 7–8, 37, 280; for the claim that non-temporal experience is
possible, see R. C. S. Walker, *Kant*, London, 1978, ch. 3, part 3, pp. 34–41;
see further below, ch. 10, section (vi).
[5] J. E. McTaggart, 'The Unreality of Time', *Mind*, 18, 1908, pp. 457–84.

equally unreal with no difference between them, there would be no grounding in reality for the direction of time. And this, as I shall argue later, is an insuperable objection. So, while acknowledging the pre-eminent importance of the present, we should, for these two reasons, reject the view that the past and the future are both equally unreal.

The remaining two views differ only with respect to the future. The most natural is that the future does not exist but the past does. We are by our own decisions in the face of other men's actions and chance circumstances weaving the web of history on the loom of natural necessity. What is already woven is part of the fabric of the universe, but what is still unwoven has as yet no substantial reality. If, however, both the past and the future exist, we should view temporal reality not as the weaving of a web, but as the unrolling of a carpet. Part is unrolled, and visible to our gaze. Part is still rolled up, and invisible to us, though not to God; but known or unknown, it is already there, and the passage of time is not a creation of new truth but an unfolding, a revelation, of antecedently existing truth.

In spite of the intuitive appeal of the former view, many thinkers have felt impelled to adopt the latter. A 'block universe' is often pictured, in which time is a dimension very much the same as the dimensions of space, and each person and thing is represented by a world-line, along which he crawls, and the events that are occurring now are just those on the particular part of the world-line he happens to have reached at the time in question. Past, present and future are all, ontologically and modally speaking, on a par, and the apparent differences between them are illusory, to be explained away in psychological terms or as a result of linguistic usage, but not to be accorded any fundamental significance. Many different arguments are adduced for this view, from the uniformity of time, from the nature of science, and in this century from the Special Theory of Relativity.

There are, as we shall see later in this chapter, deep arguments for treating all times on a par, and these, though not conclusive, have led scientists to view time as a one-dimensional continuum, with the order-type of the real numbers, which is everywhere homogeneous and isotropic. So far as science goes, it is the right view to take. It is a presupposition of scientific argument that experiments should be repeatable, and a mere difference of date should make no difference to the outcome of a scientific observation,[6] but this stipulation defines the aspects of reality that scientific investigation can discover, rather than reflects the full nature of reality itself. And although there is a respect in which all times are importantly alike, there are others in which each is significantly different.

The claim that the past and the future both exist is open to the same objection as the claim that they both do not: in according the same status to both, it fails to account for the anisotropy, that is, the inherent directedness, of time. But the time-reversibility of the classical laws of physics has led many scientists to deny that time is really anisotropic. Yet it is one of the most fundamental facts about time that the future becomes present and the present past, and not *vice versa*. Nor is this only a matter of experience, which might be explained away as a mere psychological phenomenon. The direction of time is linked with other central parts of our conceptual structure. In an earlier work I listed five: in a more recent book Horwich lists ten.[7] The conceptual cost is therefore very high. A world in which we could, like the White Queen, remember the future and could alter the past would be one in which our ordinary concepts of memory, knowledge, explanation, aspiration, ambition, action and achievement would be inapplicable. Although attempts have been made to account for these conceptual asymmetries, as well as our experience of the passing of time, by reference to the Second Law of Thermodynamics, which in turn is alleged to stem from the initial conditions that happened to obtain at the time of the Big Bang,

[6] See J. R. Lucas, *Space, Time and Causality*, Oxford, 1985, pp. 36–38, 57, 119ff.

[7] Paul Horwich, *Asymmetries in Time*, Cambridge, Mass., 1987, ch. 1, §2, pp. 4-11.

the enterprise is uncalled for and unsuccessful. Scientists have
abstracted from our ordinary concept of time a surrogate that is
not only homogeneous but isotropic, and have then sought to
put back into the new concept the directedness that they had
removed from the old. But the Second Law of Thermodynamics, though of great significance for our understanding of
time, cannot explain all time-directed phenomena – the collapse
of the wave-packet, for example – nor can itself be derived from
purely fortuitous factors obtaining at the dawn of creation.
Although there is much more to be said about the extent to
which fundamental physical laws are time-reversible,[8] the fact
that it cannot accommodate the evident directedness of time
remains an insuperable objection to the block view of the
universe.

The block view seems to be open to a further objection which
does not tell against the view that denies the existence of both
past and future. It fails to accord any special position to the
present. But that objection too is met by an appeal to the
findings of science. If science could in principle give a complete
explanation of everything, then, it would seem, everything
could in principle be predicted before it happened. Laplace
thought that, given full information about some earlier, past,
state of the universe, it would in principle be possible to predict
its exact state at every subsequent, future date.[9] Equally, it
would be possible to retrodict its exact state at any earlier date.
There would be no difference between any one date and any
other, and present, past and future would, indeed, be all alike.
Determinism would also tell against our belief that the future is
different from the past on account of the modal difference that
the future is open, something we can alter, whereas the past is
unalterable and fixed. For if determinism is true, then, granted
the initial conditions, it would be physically impossible for the
future to be any different from what it was already determined

[8] See further, J. R. Lucas and P. E. Hodgson, *Spacetime and Electromagnetism*, Oxford, 1990, ch. 7, §7.3
[9] Marquis de Laplace, *Essai Philosophique sur les Probabilitiés*, Paris, 1814,
pp. 3–4.

to be, and so the future would be no more alterable than the past. But though classical physics was determinist, quantum mechanics is not; theoreticians have laboured over many years to produce a determinist version of quantum mechanics, but they have been unsuccessful, and now, it seems, there are insuperable obstacles in principle to there ever being a quantum mechanics that is determinist.[10] Though determinism could conceivably be true, it runs against the view of physical reality suggested by modern physics and the presuppositions of rational thought – which we engage in when we try to think out whether determinism is true.[11] We are not compelled by modern science to be determinists and think that the future is exactly on a par with the past, or that the instant dividing them is like any other.

The Special Theory of Relativity has suggested that time is not an independent reality at all, but only an aspect of a four-dimensional Minkowski space-time, and some philosophers have argued further that the Lorentz transformations between different frames of reference imply that there can be no fundamental difference between future and past.[12] But the picture suggested by Minkowski space-time is a picture only, and the difference between timelike and spacelike dimensions in the Special Theory of Relativity remains deep and ineliminable. The argument that the Lorentz transformations prove that events in the past and the future are somehow simultaneous

[10] See, for example, Michael Redhead, *Incompleteness, Nonlocality, and Realism*, Oxford, 1987; Max Jammer, *The Philosophy of Quantum Mechanics*, New York, 1974; F. J. Belinfante, *A Survey of Hidden-Variable Theories*, Oxford, 1973; Bernard D'Espagnat, *The Conceptual Foundations of Quantum Mechanics*, Reading, Mass., 1976.

[11] I have argued this in detail in J. R. Lucas, *The Freedom of the Will*, Oxford, 1970. See further below, ch. 4, section (i), pp. 61–2.

[12] See, for example, Hilary Putnam, 'Time and Physical Geometry', *Journal of Philosophy*, 64, 1967, pp. 240–7; reprinted in Hilary Putnam, *Philosophical Papers*, I, Cambridge, 1975, pp. 198–205. C.Rietijk, 'A Rigorous Proof of Determinism Derived from the Special Theory of Relativity', *Philosophy of Science*, 33, 1966, pp. 341–4.

with each other is based on a simple confusion,[13] and the claim that there can be no fundamental difference between the past and the future is based on an unjustified extrapolation of the equivalence principle.[14]

The block universe gives a deeply inadequate view of time. It fails to account for the passage of time, the pre-eminence of the present, the directedness of time and the difference between future and past, and has to make out that these fundamental features of our experience and thought are merely psychological and linguistic aberrations. It also fails to accommodate the concept of agency, the belief that we can make up our minds for ourselves, and that it is up to us what we decide and what we do. And it runs counter to the thrust of quantum mechanics, the most fundamental physical theory we have, which portrays a universe that is not determinist but only probabilistic.

The view of time as a weaving rather than an unrolling is preferable from a conceptual point of view. Time is the passage from possibility through actuality to unalterable necessity. The present is the unique and essential link between the possible and the unalterably necessary. The future and the past are modally and ontologically different, and it is natural that there should be a direction from the one to the other. There is room for agency and freedom of action. The future is not already there, waiting, like a reel of film in a cinema, to be shown: it is, in part, open to our endeavours, and capable of being fashioned by our efforts into achievements, which are our own and of which we may be proud. The chance interplay of circumstance and the implementation of our designs and purposes weave together the

---

[13] See below, ch. 11, section (iv), p. 218. A useful account is given by Richard Sorabji, *Necessity, Cause and Blame*, London, 1980, ch. 6, pp. 114–19. See also K. G. Denbigh, *Three Concepts of Time*, Heidelberg, 1981, ch. 3, §6, esp. pp. 45–7; or Storrs McCall, 'Temporal Flux', *American Philosophical Quarterly*, 3, 1966, pp. 270–81; Howard Stein, 'On Einstein-Minkowski space-time', *Journal of Philosophy*, 64, 1968, pp. 5–23; John W. Lango, 'The Logic of Simultaneity', *Journal of Philosophy*, 66, 1969, pp. 340–50; Howard Stein, 'A Note on Time and Relativity Theory', *Journal of Philosophy*, 67, 1970, pp. 289–94.

[14] See below, ch. 11, section (iv), pp. 219–20.

fabric of history. Broad expresses it less metaphorically, as a continual accretion of reality: 'Nothing has happened to the present by becoming past except that fresh slices of reality have been added to the total history of the world. The past is thus as real as the present'[15] The future, however, is different. Whereas the present and past are real, the future, as long as it is still future, is not; only by becoming present is it actualised into reality: hence the passage of time and the direction of time. 'The sum total of existence is always increasing, and it is this that gives the time series a sense as well as an order.'[16] Or we may think theistically in terms of the unforgetful memory of God. In either case, the different ontological status we ascribe to the future as compared with the past will bear upon our way of talking about it, and the knowledge we can hope to have of it. If the future is characteristically open while the past is unalterable and fixed, we shall expect there to be an interplay between mood and tense such as is found in English and most European languages. Knowledge of the past, although often in practice difficult to come by, will be metaphysically straightforward, whereas those parts of the future that are as yet indeterminate not only cannot in some sense be known, but can only with difficulty be talked about. All these features are characteristic of our ways of thinking, and so support a dynamic account of time.

But there are difficulties. The concept of passage is hard to make proof against misunderstanding. McTaggart and many others following him have distinguished the 'A series' of past, present and future from the 'B series' of dates ordered by the relations of before and after, and have then concluded that the terminology of the A series, in which events start by being future, become present, and are finally past, is incoherent;[17] from which

[15] C. D. Broad, *Scientific Thought*, London, 1923, ch. 2, p. 66.
[16] Ibid., London, 1923, ch. 2, pp. 66–7; cf. p. 83. For modern expositions of this view, see W. Godfrey Smith, 'The Generality of Predictions', *American Philosophical Quarterly*, 15, 1978, pp. 15–25; and G. Lloyd, 'Time and Existence', *Philosophy*, 53, 1978, pp. 215–28.
[17] J. E. McTaggart, 'The Unreality of Time', *Mind*, 18, 1908, pp. 457–84. D. H. Mellor, *Real Time*, Cambridge, 1981.

McTaggart himself concluded that time is unreal. But most locutions can be reduced to incoherence by a sufficiently skilled and determined sceptic. In particular, if we assume that the words 'past', 'present' and 'future' are, logically speaking, predicates denoting properties of events, inconsistency follows at once. But the conclusion to be drawn is not that the A series is inconsistent, but that its key terms are not predicates, but operators.[18] Even so, difficulties remain. If they are operators, they are tense operators, and tenses, as we shall see in the next chapter, are covertly egocentric, and egocentricity has, since the time of Plato, been thought to be subjective, and not really rational or real. A tenseless view of time, in which we have dates, but no tenses or other indications of past, present, or future, has therefore considerable attractions for thinkers who distrust the personal, as tainted with subjectivity, and seek to purge objective reality of every trace of potential egocentricity.

Besides the difficulty in adequately articulating the passage of time, there is a problem of talking about what does not exist, and hence of predicting it. We seem to be committed, if the future exists, to some sort of determinism, though of a possibly unknowable kind: so, in order to escape determinist conclusions, we feel impelled to deny existence to the future, and truth to statements about it. And yet we draw back from denying all truth to predictions or speculations, or making out that the future is entirely unknowable. Aristotle could not see how to reconcile the openness of the future with two arguments, based on sound logical principles, which seemed to show that the future must be already unalterable and fixed. In chapter 9 of *De Interpretatione*, he put forward several arguments which seem to show that granted that predictions must be either true or false, the future must be already fixed and unalterable. One of his examples was tomorrow's sea battle: tomorrow there either will, or will not, be a sea battle, from which it appears to follow that it is already true today either that there is going to be a sea battle tomorrow or that there is going not to be one. He was

[18] Simon Ord, *British Journal for the Philosophy of Science*, 38, 1987, p. 134.

certain that the arguments must be fallacious, but could not make it clear exactly what the fallacy is. Throughout the Middle Ages, the Schoolmen, wrestling with the problem of reconciling God's foreknowledge with human freedom, which we shall address in chapter 3, discussed the problem of 'future contingents' as it was called; but, in spite of many useful distinctions and penetrating insights, did not clear up the confusion altogether.

Aristotle's arguments appeal explicitly to the 'Principle of Bivalence', which we shall discuss in chapter 4, and implicitly to various shifts in temporal standpoint, some mandatory, some voluntary, which underly our use of tenses. These encapsulate certain characteristic features of our thinking about time. Although in point of fact we are necessarily located in the present, in our imagination and thought we are free to adopt any temporal standpoint, past, present, or future, that we please, and view events thence. It is a deep metaphysical fact that though in our bodies we are time-bound, in our thoughts we are not. I, my mouth, my body, my hand, am imprisoned in the twentieth century. But my mind is free to range over all time. I can imagine myself having a nasty encounter with a dinosaur in the Jurassic age, viewing the battle of Marathon, listening to a speech by Pericles, addressing the United States Congress in 2020 AD, witnessing the final inauguration of peace upon earth, having difficulty in explaining myself on the Day of Judgement. I am in my mind's eye a spectator of all time, and therefore all time, when I contemplate it philosophically, must, it might seem, be on a par, every part of it possessing the same philosophical properties.

But this argument goes too far, and in making all times the same destroys the special characteristics of time, that though each temporal standpoint is accessible to me by mental projection, each is different, and offers a different perspective on events. Past pains are not present pains, nor future aspirations past achievements. Often it is of great importance, though sometimes of great difficulty, to adopt the right temporal standpoint for surveying and understanding the

course of events. The historian needs to look back on the past
not simply as deposit of unalterable fact, but as having been, in
its own time, fashioned by the actions of men. We must not
think of Wellington on the morning of June 18th, 1815, as the
man who was to win the battle of Waterloo, but as a man who
might well lose it, and for whom it was a damned close thing.
Though hindsight sometimes enables us better to see the
significance of contemporary events, it can also distort, and we
need constantly to remind ourselves that in the welter of what
was happening, the outcome was unknown and the significance
of what they were doing obscure.

The historian needs to see things not only from the right
temporal, but from the right personal, standpoint, and in moral
and political philosophy we are constantly needing to put
ourselves into somebody else's shoes or somebody else's skin in
order adequately to appreciate an action's significance. Many
philosophers have been misled to formulate the omnipersonal
perspective of moral reasoning in terms of some principle of
universalisability, which, unless very carefully explained, seems
to enjoin an entirely impersonal perspective that reduces
morality to a cold legalism and unfeeling worship of rules. In
the attempt to see things from every point of view, we are seeing
it from none. And so too with time: although I am in my mind's
eye a spectator of all time, this is not to say that all time is,
either in my view or *sub specie aeternitatis*, the same, but that
each needs to be viewed not just from my standpoint, but from
its, and in its own particularity

Tense logic is a special case of modal logic, which will be
outlined in chapter 5. The logic of temporal standpoints will be
developed in chapter 6, which together with the exegesis of
truth in chapter 4 will yield in chapter 7 a resolution of
Aristotle's puzzles. In chapters 8 and 9 a 'possible-world'
semantics for the logic of temporal standpoints is worked out.
In chapter 10 I make a tentative effort to provide a 'modal
derivation of time'. I am far from satisfied with it: the reader
may well ask whether any such project is, indeed, possible, and I
have no argument to show that it should be; but by exposing

my own efforts to criticism, I may stimulate others to do better, or demonstrate that the enterprise is either unnecessary or impossible. Finally in chapter 11 I pick up the theological implications of chapter 3 for the nature of God and the concept of eternity.

2

Tenses

(i) Conjugation

We talk about the future by means of tenses. The very word 'future' is by origin a tensed participle. Tenses are classed by linguists as deictic expressions and by philosophers as *indexical*, or *egocentric*, expressions, or as *demonstratives* or *token-reflexives*.[1] They depend for their meaning or force on the context of utterance, in much the same way as the use of personal pronouns – I, you, he, she, we, they – does. We have a systematic way of correlating utterances of the same and different sentences by the same and different people so that the same thing – the same proposition or same propositional content or the same statement – may be affirmed or denied by various people. 'I am over six foot tall' in J. R. Lucas' mouth says the same thing as 'You are over six foot tall' in the mouth of some one else addressing J. R. Lucas, or 'J. R. Lucas is over six foot tall', in the mouth of the third party, or 'He is over six foot tall' when the context makes it clear that 'he' refers to J. R. Lucas. In the same way 'There will be trouble in Ruritania in 1911' uttered before 1911 and 'There was trouble in Ruritania'

[1] See D. R. Dowty, R. E. Wall, and S. Peters, *Introduction to Montague Semantics*, Dordrecht, 1981, p. 136; the term 'indexical' was introduced by C. S. Peirce, and popularised by Y. Bar-Hillel, 'Indexical expressions', *Mind*, 63, 1954, p. 365; the term 'egocentric particular' was introduced by Bertrand Russell, *Human Knowledge*, London, 1948, ch. 4, p. 100–108; the term 'token-reflexive' was introduced by Hans Reichenbach, *Elements of Symbolic Logic*, New York, 1947, §50, pp. 284–0.

uttered after 1911 say the same thing. The test is contradiction.[2] If two utterances in their respective contexts of utterance contradict each other then the contexts are such that they are affirming and denying the same proposition; and two utterances, such that any third utterance inconsistent with the one is inconsistent with the other, express the same propositional content.

With the first and second persons and with the tensed present tense the rules are fairly specific, and the words 'I', 'you', and 'now', and the present tense carry a lot of information: but with the third person and with the past and future tenses the reference is liable to be indefinite, and we need to make it more specific by referring to the person by name or description, or by specifying the date or describing it. In such cases the person or tense becomes redundant. 'Brutus killed Caesar in 43 BC' does not depend on the context of utterance in the way that 'I went to Grantchester yesterday' does. Nevertheless something remains. My use of the third person implies that neither the first nor the second was appropriate, and my use of the past tense implies that neither the present nor the future was appropriate. *'Brutus Caesarem interfecit'* is all right in anybody's mouth except Brutus' or Caesar's or that of some one addressing either of them, when it would seem unduly impersonal and unfeeling. If a person heard me, J. R. Lucas, uttering the words 'J. R. Lucas is over six foot tall' he would assume that I, the speaker, was not J. R. Lucas, and on finding that I was might feel that I had been somewhat disingenuous, unless he realised I was a philosopher talking about indexicals. We hesitate to say that failure to use the first or second person is wrong, but it is awkward, to say the least, and often misleading. Our rules for tense are less yielding. It is positively wrong to use a past tense to refer to an event that is still to come, or to use a future of what has already happened. I cannot now say that Caesar will be killed in 43 BC, in the way that I can say 'J.R. Lucas (that's

[2] Aristotle assumes that affirmation and denial indicate sameness of topic in *De Interpretatione*, ch. 9, 18ª35–6.

me) is over six foot tall.' Typically, the tense of the sentence used is liable to alteration if utterances on different occasions are to express the same thing. In some cases – 'It is raining', 'Kenny has written another book', 'Strawson is going to publish his lecture on metaphysics' – it is clear that the tenses are conveying information: in other cases – 'William defeated Harold at Hastings in 1066', 'Britain will be short of oil by 2017', 'Mr Markwell will be Junior Proctor for 2002–3' – the tense is redundant so far as what is actually being asserted goes, although it shows something about the date of that particular utterance. If at some other time I wanted to deny what was being said, I should, if need be, alter the tenses. I might say later 'Britain *is* not short of oil in 2017' or 'Mr Markwell *was* not Junior Proctor in 2002–3', and the fact that I used these different tenses would show that I was uttering the words at a different date: but what is being expressed would depend not on the tenses used but on the dates actually specified, and *per contra*, the tense reflects not what is being said but when it is being said.

The unyieldingness of our rules for using tenses has led philosophers to seek a system of tenseless sentences for referring to dated events; Quine, in particular, has yearned for them and called them eternal sentences.[3] It is, indeed, useful to adopt such locutions as an aid to clear exposition, replacing pronouns by proper names or definite descriptions, giving explicit dates in place of all implicit or indexical indications of temporal reference, and replacing all parts of verbs by the infinitive root italicised. Thus instead of saying 'I was there', I say 'Lucas *be* in Cambridge on February 27th, 2000'; instead of 'There will be a sea battle tomorrow', I say 'There *be* a sea-battle off Naupactus on February 29th, 2000'.[4] The meaning of such locutions is clear,

[3] W. V. Quine, *Word and Object*, New York, 1960, pp. 193–4, 208, 226–7; or *Elementary Logic*, New York, 1965, p. 6.

[4] The original idea of distinguishing the tenseless present typographically is due to J. J. C. Smart, *Philosophy and Scientific Realism*, London, 1963, p. 133. He uses italicised but otherwise grammatically correct present tenses, but the grammatically incorrect use of the infinitive is clearer for reading aloud. See, more fully, Nicholas Rescher, 'On the Logic of Chronological Propositions', *Mind*, 75, 1966, pp. 75–6.

although their uncouthness should dissuade us from thinking them an improvement on ordinary language. Stronger arguments than have been adduced thus far would have to be given before we should abandon ordinary usage on platonist principle. In default of compelling reasons, it is unwise to legislate against existing usage because linguistic habits die hard, and, as we shall see, it is easy to be misled by a conflation of the new usage and the old. My use of tenseless sentences is meant only as an adjunct to, not as a general replacement of, existing usage. Sometimes total explicitness, although inelegant, is called for, and on such occasions we should be able to provide it without in any way suggesting that it should be regarded as the norm. I shall use the uncouth italicised infinitive where necessary, but continue also to use our ordinary tensed expressions, and try to understand them.

(ii) Simple Tenses

Simple tenses depend on the relation of the time of utterance to the time of the event being spoken about. Let us call the time of utterance S, the time of the event E. If S is contemporaneous with E, or roughly so, the present tense is used: if S is after E, the past: if S is before E, the future. Tenses indicate the temporal perspective of the speaker at the time of utterance towards the event. Since no man can say different things at once, the different utterances of any one speaker must be uttered at different times. But different utterances may be about the same event. I can contradict myself. I can deny at one time what I had earlier affirmed. Hence the denial and the affirmation are about the same event. And if I were to deny my denial, I should be reinstating my original affirmation, confirming it, and asserting the same thing then as I had done originally. Thus different utterances can say the same thing, but necessarily at different times, and so with different temporal perspectives being required. 'I shall be in Cambridge on the 28th' uttered on the 18th says the same thing as 'I am in Cambridge' uttered on the

28th, or 'I was in Cambridge on the 28th' uttered on the 1st of March. We can represent the three simple tenses diagrammatically thus:

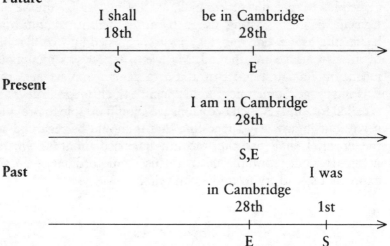

Future

| I shall | be in Cambridge |
| 18th | 28th |

   S   E

Present

I am in Cambridge
28th

S,E

Past

         I was
in Cambridge
28th   1st

   E   S

(iii) Reichenbach's Reference Point

Our account of tenses thus seeks to accommodate the differing temporal perspectives the speaker may have on the same event on different occasions of saying the same thing about it. But a more complicated analysis is needed. For what will be in the future is not only as of now future, but is itself a temporal point of view which will in due course be the actual, present point of view. All our tomorrows will, in their day, be called 'today', as were all our yesterdays in their day too. And we know this. Although our actual temporal position is necessarily what it is, we know that there *be* other ones, and can envisage our occupying them, and viewing events from that temporal standpoint.[5] Just as when I am working out the logic of first-, second- and third-persons, I have to remember that you use the

[5] See above, ch. 1, pp. 11–12.

word 'I' of you, as also he does of him and she does of her, so
we must all remember when working out the logic of tenses that
now is not the only pebble among the sands of time, even
though it is the one that is currently our vantage point.

Many tenses are, in consequence, more complex than the
simple account allows, and require a more complicated
analysis. Reichenbach[6] argues that we need always to distinguish not only the date of the event, E, and the date of
utterance, S, but a third date, which he calls the reference point,
R. The reference point reflects the ability of the mind to project
itself to any time, and to see things from any temporal
perspective. Although I can only utter a sentence at a time at
which I happen to be, and although I must, if I am to succeed in
saying anything relevant, give events their appropriate dates, I
can view these events from any temporal point of view I care to
adopt. Thus I can view a past event as being antecedent to a
reference point itself earlier than the date of utterance, in which
case I use the pluperfect tense – 'when I saw you last week I had
not heard of your new appointment',

Pluperfect

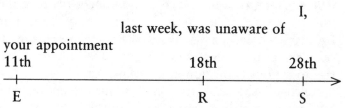

which we can symbolize more shortly by (E–R–S). Similarly,
Reichenbach analyses the future perfect thus (S–E–R):

Future perfect

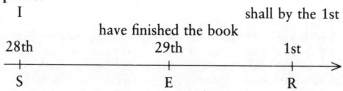

[6] Reichenbach, *Elements of Symbolic Logic*, pp. 287–96.

Reichenbach's analysis also enables us to give an outline account of the difference between the aorist and perfect tenses. The aorist tense (or simple past, as Reichenbach calls it), 'I was in Cambridge on the 27th', views my being there from a reference point contemporaneous with the event,

Aorist

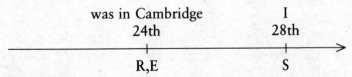

whereas the perfect tense, 'I have been in Cambridge' views it from a reference point contemporaneous with my utterance.

Perfect

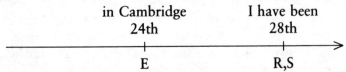

If we consider that any of the three variables, S,R and E, may be either before, or after, or contemporaneous with, either of the others, we see that there will be a very large number of Reichenbach tenses. Many of these are not distinguished in ordinary usage, sometimes for good reason, sometimes giving rise to serious confusion in consequence.

Reichenbach's introduction of a third temporal reference in addition to the date of the event and the date of utterance, is important and controversial.[7] It is the linguistic expression of the metaphysical fact, noted in the previous chapter, that although we are timebound, our thoughts are not so. If I am to

---

[7] Grammarians talk of the aspect of tenses, which is not deictic, but differs from Reichenbach's reference point in not being primarily matter of temporal standpoint, but of whether the verb is either perfective or not, or progressive or not; see Randolph Quirk, Sidney Greenbaum, Geoffrey Leech and Jan Svartvik, *A Grammar of Contemporary English*, London, 1972, 3.36–3.42, pp. 90–7; or their *A Comprehensive Grammar of the English Language*, London, 1985, 4.17ff., pp. 188ff.

say anything, I must say it now, in the time of my corporeal existence. But not only can I talk about events at other times than my own, I can also talk about them from other temporal standpoints than my own present one. Although I have to say whatever I have to say, at the time and place where I am, February 28th, 2000, in Cambridge, what I have to say not only may be about other events at other times and other places, but also viewed from yet other times and other places again. Much as I can view Cambridge from the Gogs, from Madingley Hill, or from the A603, so I can also view Cambridge events as a schoolboy seeking admission, as a graduand receiving his degree in the Senate House, as an old member coming up for his college gaudy, or as an old man reminiscing to his great-nephew. I am not confined to the here and now in my mind as I am in my body, and can think, and so also talk, of myself anywhere and anywhen, and therefore as viewing places and events anywhence and from any temporal standpoint.

If we allow such a liberty, we must allow its extension to iterated perspectives. Not only can I think of myself as an old man reminiscing to his great-nephew about his Cambridge days, but I can think of myself as an old man reminiscing to his great-nephew about how he looked forward to Cambridge when he first came up for admission. We can iterate reference points, and distinguish not just one R, but an R_1, R_2, *etc.*, and on one occasion it will be crucial to the argument.[8]

Even if we allow the possibility of reference points, we may question their necessity. The simple analysis of the simple tenses given above has appeared adequate to most thinkers, and it seems a needless complication to insist on a gratuitous reference point when none is really needed.[9] Nevertheless it is wise to insist. If we do not, there is a danger of a double assimilation, with the simple past, for example, (E–S) being sometimes construed as a perfect, (E–R,S), and sometimes as an aorist (R,E–S). In simple cases that may not matter; in American

[8] See below, ch. 7, §(i), p. 121.
[9] See further below, ch. 6, §(i), p. 110.

English and in modern French the perfect is used indifferently for either tense: but in complicated arguments in which there are significant shifts of temporal perspective, the danger of equivocation is real.

Reichenbach's analysis has its inadequacies, especially with regard to the perfect, as he himself admits.[10] He points out that besides speaking of events that occur more or less instantaneously, we speak also of processes and activities that go on for some time: I was enjoying myself, I am writing a book, I shall be having lunch. The distinction between instants and intervals is a difficult one mathematically, and has given rise to many puzzles about the nature of time, in particular St Augustine's paradox of the vanishing present.[11] But they are not germane to our present purpose, and though Reichenbach's analysis needs further refinement, it is nonetheless illuminating, enabling us to make crucial distinctions, and going far to explain our rules for the sequence of tenses.

Reichenbach's analysis of tense is illuminating because it shows that tenses perform two different functions: they show the relation of the speaker to what he is talking about; and they view events from a particular time. The former must vary from one occasion of utterance to another, but the latter is a matter of choice, and gives rise to complicated rules, both as regards 'sequence of tenses' and as regards entailments and truth. It is

[10] For further refinements, see Anthony Kenny, *Act, Emotion and Will*, London, 1963, ch. 8, pp. 273–5. For more recent accounts and criticisms, see B. Taylor 'Tense and Continuity', *Linguistics and Philosophy*, 1.2, 1977, pp. 199–220; C. S. Smith, 'The Syntax and Interpretation. of Temporal Expressions in English', *Linguistics and Philosophy*, 2.1, 1978, pp. 43–100; C. S. Smith, 'Constraints on Temporal Anaphora', *Texas Linguistic Forum*, 10, 1978, pp. 76–94; R. W. McCoard, *The English Perfect Tense–Choice and Pragmatic Inferences*, Amsterdam, 1978, esp. pp. 88, 195; Lennart Aqvist, 'A Conjectured Axiomatization of Two Dimensional Reichenbachian Tense Logic', *Journal of Philosophical Logic*, 8, 1979, pp. 1–45. For important differences between the aorist and perfect noted by Aristotle, see Richard Sorabji, *Time, Creation and the Continuum*, London, 1983, ch. 1, pp. 10–12.

[11] See further, J. R.Lucas, *A Treatise on Time and Space*, London, 1973, §4, pp. 20–5.

again helpful to compare conjugation by tenses with conjugation by persons. The speaker uses, and is required by the rules of grammar to use, the first person of himself and the second of the person he is talking to. But he does not have to consider everything from his own point of view. I can see things your way, and I can stand up for an absent friend. A barrister uses the first person to refer to himself, but represents his client's interests and argues his side of the case. We recognise that different people have different opinions, different desires, different inclinations, and generally different points of view, and can within limits enter into those of other men even though we do not hold them ourselves, just as we can talk today of what I shall have done next year, or what I was going to do last week. To assess your interests or your rights demands the same detachment from my own situation as viewing the year's work from this time next year, or the week's activities from this time last week. In the one case I have to distance myself from my interests or rights, in the other from from the present time at which I am actually talking. And as I may have to weigh the different interests different people may have in some project – that an airport be built in a particular locality or that the family should go for a skiing holiday instead of spending it by the sea – so I may have to view the same event from different temporal points of view, and consider how its truth from one of them bears on its truth at another.

We thus have two different changes of temporal standpoint to consider: the change of the date of utterance, S, which happens irrespective of our wishes by the mere effluxion of time; and the change of the reference point, R, which is at our discretion. These changes are independent of each other. Thus in the simple case where speaking of my being in Cambridge on the 28th, I speak on the 18th, the 28th and the 29th respectively, I use the locutions 'I shall be in Cambridge', 'I am in Cambridge' and 'I was in Cambridge'. The changes are due to the changing temporal positions of the date of utterance. Similarly if I am viewing my being in Cambridge on the 28th from the temporal standpoint of the 29th, on the 18th I say 'I shall have been in Cambridge', on the 29th I say 'I have been in Cambridge', and on the 1st of March I say 'I had been in Cambridge', as in, for example, 'Since I had been in Cambridge, I was able to tell them what was going on in the English faculty'. A change in the date of utterance requires a suitable change in tense, but does not alter the temporal standpoint which we have *voluntarily* taken up. Essentially, the temporal perspective of Reichenbach's reference point is part of the invariant core of what may be asserted or denied on different occasions, in contrast to that of the date of utterance, which is not part of what is asserted or denied but only indicates the appropriate stance to be taken towards it at the moment of utterance. We are thus led to consider, as the topic of tense logic, not just simple propositional contents, but also propositional contents viewed in a temporal perspective from a particular point of view; not just the fact that Lucas *be* in Cambridge on the 27th, but that fact as seen from the point of view of the 28th. If we have a dated propositional content, say p_{27}, we can express the more complicated invariant core, which may be asserted or denied by differently tensed sentences uttered on different occasions, by enclosing the propositional content in square brackets, and giving the date of the reference point as a subsequent subscript, thus

$$[p_{27}]_{28}$$

It is 'RE' propositions of this form that are on different occasions, and so by means of different tenses, affirmed or denied, thus constituting the invariant propositional core of our tensed discourse.

Logic traditionally has been concerned with entailment patterns, which depend on the connotations of words, and so are manifested by sentences that are independent of context rather than utterances whose force largely depends on their context.[12]

---

[12] For further brief discussion of utterances, sentences and propositions, see below, ch. 4, §(ii), pp. 62–5; for the distinction between particular propositions, p_{28}, and propositional variables, p, p_t, *etc.*, see below, ch. 5, §(i), p. 80.

Hence, it is easier to consider the logical relations between different RE propositions, each one the invariant core of different utterances tensed appropriately for their different occasions of utterance, than to go deeply into the way in which the expression of an RE proposition varies with time. The former will be discussed in chapter 6, the latter will begin to emerge only in chapters 8, 9 and 10.

(iv) Future Forms

It is always illuminating to hold a mirror up to time, and examine its mirror image, in which past and future are interchanged. Reichenbach's distinction between the aorist and the perfect prompts us to ask what the mirror image future tenses are. We find that in most languages there are no clearly distinguished analogues of the two past tenses. The two forms, which Reichenbach calls the simple future and the posterior present respectively,

$$—S—R,E \rightarrow \qquad \text{and} \qquad —S,R—E \rightarrow$$

F(i)	F(ii)
Simple future	Posterior present

are both normally rendered by the simple future 'There will be a sea battle tomorrow'. This distinction is crucial. There is a very great difference between on the one hand the simple future, F(i), which is the future analogue of the aorist in English and Greek, and on the other the posterior present, F(ii), which is the future analogue of the familiar perfect tense. The essential difference is that the simple future speaks only about tomorrow, that it is a sea-battle day, whereas the posterior present says something about today too, that it is a day-before-a-sea-battleday, just as when I use the simple aorist 'I was in Cambridge yesterday' I am only talking of yesterday, namely that it was spent by me in Cambridge, whereas if I use the perfect 'I have been to Cambridge' I am talking also about the present-day situation – e.g. that I know what King's College Chapel looks like, or that I do not need to go there again immediately. Similarly we distinguish what we say about the day before Pompeii was destroyed – which we should describe as being much like any other day and not at all pregnant with disaster – from what we say about the day before Byzantium fell – which could not be adequately described except as the eve of defeat. And, conversely we distinguish the day of preparation for the sea battle with the oarsmen practising their strokes and the helmsman making sure they can keep in line, from the day which we do not wish to describe as a pre-sea-battle day, but rather as one on which, as usual, the Greeks combed their flowing locks and ran races on the sand.

In order to represent Reichenbach's distinction between simple future and posterior present I shall use the ordinary 'shall' and 'will' for the simple future, and render the posterior present, with some infelicity, by 'is going to',[13] – 'There is going to be seabattle tomorrow' – and the corresponding *je vais* in French or μέλλω (*mello*) in Greek plus the infinitive, and the periphrastic future in Latin and Greek – *pugnaturi sumus cras*, μαχούμενοί ἐσμεν αὔριον (*machoumenoi esmen aurion*). This usage is schematic, and on some occasions awkward,[14] but I shall use it none the less as a term of art to represent as colloquially as I can Reichenbach's posterior present.

The use of the present-tense form 'is going to' is significant. It is as though we were linguistically uncomfortable with pure futurity, and did not know how to refer to what was not yet in existence, and preferred to talk of present tendencies instead, which could be referred to properly, and would in due course issue in the future event we wanted to talk about. In many other Indo-European and Semitic languages the future tenses are defective in comparison with the present and the past. Often there is no future participle or infinitive (as in French) or no future subjunctive (as in Latin and Greek), and in most modern

---

[13] Quirk et al., *A Grammar of Contemporary English*, 3.29, pp. 87–8; or their *Comprehensive Grammar of the English Language*, 4.43 pp. 214–15.

[14] See below, ch. 6, §(ii), p. 118, and ch. 7, §(i), pp. 123–4.

languages the future is awkwardly constructed from some not wholly appropriate auxiliary verb.

English is particularly awkward. The first person, according to the grammar books, is formed with 'shall', the second and third persons with 'will'. In Scotland, and often in England, however, 'will' is used in all three persons, and in the corresponding conditional mood it is sometimes unnatural to say 'I should' rather than 'I would'. We can, none the less, give some rationale for the conjugation 'I shall', 'you will', 'he will', in terms of our different access to our own and other people's minds. The word 'shall' has a modal force, like the German *sollen*, which survives in 'you shall' and 'you should'. My future actions are tied to my moral obligations and practical necessities as perceived by me. I am privy to my own deliberations, and know what calls morality, prudence and pleasure are making, and although I may, under the influence of passion, misjudge where the balance lies, I do not at the time in question recognise the mistake, and act on what I, perhaps mistakenly, take to be the correct conclusion of my deliberations. In my own eyes at the time in question I am usually doing what I think I should be doing; I so much think of myself as doing what I should be doing, that my future actions can be represented by me as conforming to what I have got to do. 'I have got to be going' or 'I must go now' says the guest announcing his impending departure to his host: it is not what he wants to do, but what, in view of other obligations and calls on his time, he reckons, all things considered, he must do. With other people, however, I am less sure of the conformity of their actions to the calls of duty or prudence. I may know what you should do, and may tell you, but it would be rash to assume that you will act on it. The word 'will', like the Latin *volo* and its cognates, is equivocal between inclination and determination. It can mean what one would like to do, and would do readily and willingly in the absence of considerations to the contrary. It can mean also what one is determined to do, having made up one's mind by an act of will, as when bride and bridegroom publicly avow their decision to take each other in wedlock. The difference between inclination and determination is an important one of which we are well aware in our own lives, but when I come to ascribe states of mind to other people, the non-linguistic behaviour which warrants my believing that some one wants to do something is the same as that which warrants my believing that he has decided to do it. Not having private access to his mind, I am not privy to the deliberations which make the difference between the *prima facie* propensities, or inclinations, which he takes into account as he tries to make up his mind, and the conclusion of the debate, when he has made up his mind and decided what he is going to do. Unless he makes me his confidant, which cannot be very often, I cannot tell whether his present tendencies, on the basis of which I predict his future actions, constitute for him the *agenda* or the *acta* of the debate, his first words or his last. All I can tell from the outside is that he is manifesting certain tendencies or propensities on the basis of which I predict his future action, and say that he will do it. If all I have to go on is overt behaviour, without any linguistic avowals, then although I can draw the distinction in my own case between what I want to do and what I am resolved to do, the evidence available to me in his case does not differentiate between them, and one word serves for both, and does duty also for expressing futurity.

Other languages do not differentiate between the persons. French is like the English first-person form – *je mangerai, tu mangeras, il mangera*, I have to eat, you have to eat he has to eat, express futurity by a modal concept. Modern Greek – θὰ (*tha*) – is like the English second- and third-person form. The German *werde* has no sense of personal obligation, inclination or decision, but is an expression of present process, and thus again expresses futurity in terms of present tendency, though in less modal form.

Logic is thin, and difficult to grasp. We find it much easier to feel our way into problems of temporal logic and future truth if we consider first the more substantial question of God's foreknowledge of the free actions of human agents. The problem seems much more real in its theological form. Few people are greatly bothered by logical fatalism as Aristotle or subsequent logicians have expounded it. There must be a fallacy somewhere, we think, and if we had time, it might be interesting to track it down, but the outcome is not seriously in doubt. Whereas with God, many, many men have been deeply disturbed by the thought that He might know already what we are going to do, and have cudgelled their brains to find some way out. Theological determinism is far easier to understand, its problems are more live, its solutions more widely canvassed and more closely scrutinised. And so to theological determinism we now turn.

3

Foreknowledge and Freedom

(i) The Problem

If God knows what I am going to do tomorrow, then I cannot help doing it. For if when tomorrow comes I do not do it, then God will not have known what I was going to do, but only had a mistaken belief about it. Therefore either God did not know, or I cannot help doing actions many of which seem to be free actions. If the former, God seems much less omniscient than we might reasonably expect, and open to the jibes of some latter-day Elijah or Isaiah: if the latter, free will is an illusion, and all our concepts of responsibility and desert misplaced. We seem to be called upon to abridge either the perfection of God or the autonomy of man, and whichever horn of the dilemma we choose to impale ourselves on, we are left with something very different from the Christian religion, which affirms both the greatness of God and the freedom and moral responsibility of man.

Many thinkers have chosen the latter horn of the dilemma, and denied that responsibility requires a real possibility of doing otherwise. It is only pride, they say, that makes man think he is free. In truth he is utterly dependent on his Creator, and can do nothing, achieve nothing, attempt nothing, without God. We have no power of ourselves to help ourselves, and should ascribe all might, majesty, dominion and power to God, who is the author of all things, and orders all things according to the good purposes of His perfect will. But this, though a possible form of theism and couched in traditional language, runs counter to the whole tenor of the Christian religion. It denies in its extreme theocentricity the humanity of man. And man, according to the Old Testament, was made in the image of God, and, according to the New, is what God became. Christianity is not simply theism, but is theism with a human face. To deny man's freedom of will is to deny an essential attribute of humanity, and to undercut the specifically Christian attributes of the Christian religion.

To deny man's freedom of will also raises the problem of evil in an insoluble form. For in ascribing all power to God, it ascribes to Him all responsibility for the course of events in the created world, which in spite of its many excellences and beauties is also for many a vale of tears. Although the 'Free-Will defence', as Alvin Plantinga calls it,[1] is not by itself enough to explain all the evil in the world, it goes a long way to resolving the problem of evil at the most difficult point. Many of the worst evils are the ones intentionally inflicted. If we deny the freedom of the will, we cannot avoid holding God responsible for the evil which wicked men decide to do. He could have made them decide otherwise. God could have made Stalin and Hitler, Obote and Amin, turn away from their fiendish cruelties and walk in the ways of gentleness and peace.

Most Christians have, therefore, been unwilling to abandon freedom, and have sought in some way or other to reconcile God's foreknowledge with human freedom. Many thinkers have argued that there is no incompatibility because 'God's foreknowledge is not the cause of things future, but their being

---

[1] Alvin Plantinga, *The Nature of Necessity*, Oxford, 1974; *God, Freedom and Evil*, London, 1975. See earlier, Plato, *Republic* bk X, 617e, αἰτία ἑλομένου· θεὸς ἀναίτιος (*aitia helomenou: theos anaitios*), the blame for evil lies on the chooser not on God; and St Augustine, *Retractationes*, bk I, ix, 1, *Malum non exortum nisi ex libero voluntatis arbitrio*. Evil has arisen only from the free choice of will. See more fully, St Augustine, *De Libero Arbitrio*, bk I, i, 1. English translation in J. H. S. Burleigh, *Library of Christian Classics*, vol. VI, London, 1953, p. 113; and Methodius, *De Autex*,(in J-P. Migne, *Patrologia Graeca*, Paris, 1837, 18, 254f., tr. in G. L. Prestige, *God in Patristic Thought*, London, 1936, pp. 28ff.

future is the cause of God's foreknowledge that they will be'.[2] Milton uses this argument to acquit God of responsibility for man's misdeeds:

They, therefore, as to right belonged,
So were created, nor can justly accuse
Their Maker, or their making, or their fate,
As if Predestination overruled
Their will, disposed by absolute decree
Or high foreknowledge; they themselves decreed
Their own revolt, not I: if I foreknew,
Foreknowledge had no influence on their fault,
Which had no less proved certain unforeknown.[3]

Samuel Clarke argues in similar vein:

... Fore-knowledge in God can be consistent with Liberty of Action in Men ... because Fore-knowledge has no Influence at all on the Things Fore-known; and it has therefore no influence upon them, because Things would be just as they Are, and no otherwise tho' there was no Fore-knowledge. Fore-knowledge, does not cause things to be; but Things that are to be hereafter, whether necessarily or freely, are the cause of its being fore-known that they shall so, whether necessarily or freely, be. The Futurity of free Actions is exactly the same, and, in the nature of things themselves, of the like certainty in Event, whether they can, or could not be, fore-known. And as Our knowing a thing to be, when we see it is; does not hinder an Action from being Free, notwithstanding that it is then certain and cannot but be, when it is: so God's foreseeing, that any Action will freely be done; does not at all hinder its being Free, tho' he

---

[2] Origen, *apud Eusebius, Praeparatio Evangelica*, bk 6, ch. 11, pp. 286, 287, in Migne, *Patrologia Graeca* 21, 492; excerpted and quoted by Dr Daniel Whitby, *Discourse on the Five Points*, Discourse VI, London, 1710, ch. 1, 1, pp. 494–5, 2nd ed London, 1735, p. 473. Compare Origen, *De Oratore* VI, 4, tr. J. L. E.Oulton and H. Chadwick, *Library of Christian Fathers*, vol. II, London, 1954, p. 252, 'but the foreknowledge of God is not the cause of all future events'; and St Augustine, *De Libero Arbitrio*, III, iv, 10, tr. Burleigh, *Library of Christian Classics*, p. 177, 'God compels no man to sin, though he sees beforehand those who are going to sin by their own will.'

[3] J. Milton, *Paradise Lost* III, ll. 111–19; my italics.

knows it certainly; because His Fore-seeing things to come, does not more influence or alter the Nature of things, than our seeing them when they are[4].

A similar argument has been more recently put forward by Linda Zagzebski, who gives a definition of causal contingency in terms of possible worlds such that our future actions may be causally contingent even though foreknown by God. Hence God's foreknowledge does not imply causal determinism, and, it is argued, there is no incompatibility between divine omniscience and human freedom.[5]

But these arguments miss the point. They show that men's actions are not causally necessitated by God's foreknowing them, but not that they are not necessitated at all. Jonathan Edwards puts the point clearly, as did Boethius before him.[6]

To all which I would say; that what is said about knowledge, its not having influence on the thing known to make it necessary, is nothing to the purpose, nor does it in the least affect the foregoing reasoning. Whether prescience be the thing that *makes* the event necessary or no, it alters not the case. Infallible foreknowledge may *prove* the necessity of the event foreknown, and yet not be the thing which *causes* the necessity. If the foreknowledge be absolute, this *proves* the event known to be necessary, or proves that 'tis impossible but that the event should be, by some means or other, either by a decree, or some other way, if there be any other way: because, as was said before, 'tis absurd to say, that a proposition is known to be certainly and infallibly true, which may possibly prove not true.

The whole of the seeming force of this evasion lies in this; that, in as much as certain foreknowledge don't *cause* an event to be necessary, as a decree does; therefore it don't prove it to be necessary, as a decree does. But there is no force in this arguing. For it is built wholly on this supposition, that nothing can *prove*, or be an *evidence* of a thing's being necessary, but that which has a *causal influence to make it so*.

---

[4] Samuel Clarke, Sermon XI, 'Of the Omniscience of God', reprinted in *Works of Samuel Clarke*, London, 1738, vol.I, pp. 69–70.

[5] Linda Zagzebski, 'Divine Foreknowledge and Human Free Will', *Religious Studies*, 21, 1985, pp. 279–98.

[6] *De Consolatione Philosophiae*, V, iv, ll. 27ff.

But this can never be maintained. If certain foreknowledge of the future existing of an event, be not the thing which first makes it impossible that it should fail of existence; yet, it may, and certainly does *demonstrate*, that it is impossible it should fail of it, however that impossibility comes. If foreknowledge be not the cause, but the effect of this impossibility, it may prove that there is such an impossibility, as much as if it were the cause. It is as strong arguing from the effect to the cause, as from the cause to the effect. 'Tis enough, that an existence which is infallibly foreknown, cannot fail, whether that impossibility arises from the foreknowledge or is prior to it. 'Tis as evident, as 'tis possible should be, that it is impossible a thing which is infallibly known to be true, should prove not to be true; therefore there is a *necessity* that it should not be otherwise; whether the knowledge be the cause of this necessity, or the necessity the cause of the knowledge.[7]

The crucial argument is that God's foreknowledge, even though it may not cause a man's future actions, nevertheless precludes their being the result of his subsequent free choice. Foreknowledge is not the same as predestination, but if it means that our future actions are already no longer capable of being different, then it is equally incompatible with free will.

(ii) Knowledge and Necessity

Although the contention *If God knows what I shall do tomorrow, then I cannot do other than do it* has intuitive plausibility, it is difficult to formulate precisely the argument for it, and many philosophers have suspected that it rests upon a muddle. Boethius put forward the suggestion, which he did not entirely accept himself, that in the equivalent thesis *If God knows what I shall do tomorrow, then I necessarily shall do it* there is an ambiguity in the scope of the *necessarily*. We can express the relation between the two sorts of necessity by the use of brackets. Should we construe

[7] Jonathan Edwards, *Freedom of the Will*, part II, section XII, pp. 123–24, in 1969 edition, ed. A. S. Kaufman and W. S. Frankena; pp. 173–5 in 3rd ed. London, 1760; pp. 179–81, London, 1831; pp. 176–7 Liverpool, 1877.

If God knows what I shall do tomorrow, then I necessarily shall do it

as

necessarily (If God knows what I shall do tomorrow, then I shall do it)

or as

If God knows what I shall do tomorrow, then (necessarily I shall do it)?

or, more idiomatically, as, on the one hand,

It is necessary that (If God knows what I shall do tomorrow, then I shall do it)

or, on the other,

If God knows what I shall do tomorrow, then (it is necessary that I shall do it)?

Formal logic enables us to put the distinction more succinctly. We symbolize *if . . . then . . .* by →, and necessity by a box symbol, □ or by a bold capital **L**,[8] and then express them as

$$\Box(p \rightarrow q), \text{ and } p \rightarrow (\Box q) \text{ respectively,}$$

though normally we write the latter omitting the brackets:

$$p \rightarrow \Box q.$$

Later the Schoolmen distinguished the two by using the term *necessitas consequentiAE* of the former,

$$\Box(p \rightarrow q)$$

and *necessitas consequentIS* of the latter,

$$p \rightarrow \Box q.$$

With the aid of this distinction we can see how there might be a fallacy in the argument from God's foreknowledge to the

[8] For full list of the symbols of formal logic, see below, ch. 5, p. 81, n. 2.

conclusion that we have to do whatever God knows that we shall do. It might be that we could prove the *necessitas consequentiae* It is necessary that (*If God knows what I shall do tomorrow, then I shall do it*), which in ordinary English could be expressed as *If God knows what I shall do tomorrow, then necessarily I shall do it*, which in turn could be understood as a *necessitas consequentis*, that is, as meaning *If God knows what I shall do tomorrow, then (it is necessary that I shall do it)*.

Certainly there are arguments, which we shall develop in section (iv), from the grammar of the word 'know' in favour of the former proposition, It is necessary that (If God knows what I shall do tomorrow, then I shall do it), and which give us good reason to suppose that if something is known, it necessarily is true in the one sense. Hence there is at least the possibility of a fallacious argument from the one to the other. And no doubt some people have argued fallaciously. But in order to show that the argument from foreknowledge is in general fallacious, it is required to establish that the only available argument rests upon an equivocation between a *necessitas consequentiae* and a *necessitas consequentis*, and that on the one hand it is not possible to argue for a *necessitas consequentis* directly, and that on the other there is no sound argument from foreknowledge solely in terms of a *necessitas consequentiae*. In fact, neither of these conditions holds. There are two separate arguments, both weighty, one involving a *necessitas consequentis*, the other a *necessitas consequentiae*, and much of the difficulty we have in giving an adequate account of foreknowledge generally, or in reconciling it with human freedom in particular, arises from there being two, separate but easily confused, arguments in the field at the same time.

(iii) *Necessitas Consequentis*

Many philosophers have thought that the meaning of the word 'know' is such that all knowledge is necessary. Consider the two locutions:

    a) I think that p, but it may not be true that p.
    b) I know that p, but it may not be true that p.

The former is a perfectly possible locution, which we often have occasion to use – for example, I may say 'I think there is a train at 10.22, but that may not be true' – whereas the latter is not acceptable; if I allow that it may be false, then I have no business to maintain that I know it. Any person who uttered the words 'I know it, but it may not be true', would thereby show that he did not understand the rules of the use of the English words 'I', 'know', 'may', 'not' and 'true'. Plato maintains, therefore, that if a thing is to be known, it must be not only true, but infallibly true. And in that case, *If God knows that I shall go to Professor Strawson's lecture tomorrow*, then it follows that *it is necessary that I shall go*, or, more colloquially, that *I must go to Professor Strawson's lecture tomorrow*.

    Plato has had many followers. Descartes, Locke, and in our own time A. J. Ayer, have all been telling us that we do not really know – 'philosophically know', as we might term it – all sorts of things we thought we did, but only probably believe them. Most disciplines that are normally accounted knowledge are by this criterion excluded from being real knowledge. All human affairs, which Aristotle characterized as being 'capable of being otherwise' τὰ ἐνδεχόμενα ἄλλως ἔχειν (*ta endechomena allos echein*) lack the logical necessity that Plato sought. The natural sciences are in better case, but even they lack logical necessity, and, dealing with the transient phenomena of this fleeting world, cannot acquire the absolute certainty that Plato regards as the true mark of knowledge. Only logic and mathematics deal with truths that are logically necessary and therefore only these are to be accounted knowledge in the Platonic scheme of things. Other philosophers are less stringent, and admit, besides the *a priori* timeless truths of mathematics, the omnitemporal truths of the physical sciences; others allow particular present truths given immediately by sense experience, and past truths, which are unalterable and so, in a different sense of the word 'necessary',

necessary. But all agree that only rather high-grade truths, which are in some sense necessarily true, can be the proper object of knowledge, so that if God knows what I am going to do, it is necessary that I do it.

There must be something wrong with this argument. It is not how we normally use the word 'know'. We do not normally refuse to say that we know that there is a train at 10.22, or that Queen Elizabeth II is our queen, on the grounds that these are not necessary truths, and might conceivably be false. Although there is a sense in which if I know that *p*, then *p* must necessarily be true, the necessity is not a logical necessity. Whereas it is a logical necessity that *If someone knows that* p *then* p *is true*, it is not the case, in the ordinary way of speaking that *If someone knows that* p, *then it is logically necessary that* p. It is perfectly possible for someone to say that he knows that *p*, though *p* is not logically necessary, that is to say, it would not be inconsistent to deny *p*. The modal word 'necessary', along with its correlates, 'possible' and 'impossible', and equivalent terms, such as 'must', 'can', 'cannot', and the like, apply in many different modes.[9] Besides logical necessity, there are physical necessity, biological necessity, social necessity, moral necessity, and many others, and likewise physical possibility, physical impossibility, biological possibility, and so on. Where Plato and his followers have erred was in construing the modality of the *may be* as being that of logical possibility rather than some other, more work-a-day modality. Although it is wrong to say 'I know that *p*, but *p* may be false', it is not wrong to say 'I know that *p*, though *p* might just conceivably be false – it is logically possible that it be false'. Thus it is that we often say we know things that could conceivably be false. I know that there is a train at 10.22, I say, though it is logically possible/physically possible/biologically possible/legally possible that there has been a derailment, or a lightning strike by the engine-drivers. I know that Elizabeth II is our Queen, though it is possible that

---

[9] See further below, ch. 5, §(iii), p. 92; ch. 7, §(iii), pp. 130–1; ch. 8, §(i), p. 138.

she has died in the last few minutes, and that Charles III is now our lawful King. What Plato, Descartes, Locke and other philosophers have done is to apply the wrong modality to their exegesis of knowledge, so as to make it almost impossible for anyone to know anything. And what we have to do in consequence is not to deny the exegesis, but to make explicit the modality involved.

Ordinary knowledge is, *pace* the philosophers, not logically guaranteed to be infallible. In order to be able to claim knowledge, I need to have adequate reasons for making my claim and not to have any substantial reason for doubting it, but these reasons do not have to be entailments of deductive logic. What counts as an adequate reason or substantial doubt varies with the context. Different contexts construe 'may be false' with different modal force. An astronomer can properly claim to know when there will be an eclipse of the sun, because that is what is predicted by informed and competent astronomical calculations, even though a wandering comet or black hole might so affect the revolution of the earth or the orbit of the moon that the prediction turned out in the event to be mistaken. You can properly say that you know I shall go to Strawson's lecture tomorrow if I have told you that I shall go, though it is possible that I might be run over by a bus on my way there, or I might have become a Deconstructionist, and set off for Paris to begin a new, more authentic, life there as an art critic or novelist, or decided to go gay in San Francisco. If this were to happen, you would have to withdraw your previous claim to knowledge, and say 'I thought I knew, but I was wrong: he did not turn up but had gone mooching off to Paris.' Of course, you should not say that you know unless you have good grounds for claiming knowledge. If you have less than adequate grounds, or if you have some reason for doubting, then you ought not to give an unqualified guarantee, but hedge, and say only 'I think' rather than 'I know'. But in the context of human affairs, it is quite proper to say 'I know' on the strength of a reliable avowal of intention even though accident or infirmity of purpose may intervene to prevent good intentions issuing in action. The modality is that of human affairs, which is different from that of physical necessity, which is different in turn from that of logical necessity. Different modes of discourse have different modalities, and a type of possibility that would invalidate a knowledge claim in one mode of discourse may not rate as a substantial doubt in another.

We thus have found a cogent argument, based on the grammar of the word 'know' for the contention that if God – or anyone else – knows that *p*, then it is necessary that *p*: but the necessity involved varies, and is not confined to logical necessity, nor to physical necessity, nor to any other threatening sort, but may be some necessity that does not preclude a subsequent, though entirely unexpected, change of mind. If I were to become an Deconstructionist in Paris, or go gay in San Francisco, you would, I hope, be very surprised, but would not have to revise the laws of physics, in the way you would if an expected eclipse failed to occur. It would be simply another instance of the frailty of human nature, something you were sadly aware of already, though trusting not to find it exemplified in this instance. And if, on the other hand, I continue not beating my wife, keeping promises, telling the truth, turning up for lectures, and not maximising the misery of mankind, the fact that you had predicted it all does not derogate from my freedom. For the modalities involved were internal reasons not external causes, reasons which it was up to me whether to adopt or not, rather than causes which would operate on me willy-nilly. Foreknowledge may involve modality, but some modalities do not foreclose freedom.

(iv) *Necessitas Consequentiae*

The contention that *It is necessary that* (*If God knows what I shall do tomorrow, then I shall do it*) can also be based on the grammar of the word 'know' It stems from the linguistic fact that it is inconsistent to say of anyone

He knows that s, but s is not true,

or

He knows that s, but not-s is true

which we can symbolize

$$Gks \ \& \ -s \vdash^{10}$$

from which it follows that *Gks* entails *s*

$$Gks \vdash s,$$

and that it is a tautology that *Gks* implies *s*, which we can formalise by

$$\vdash Gks \rightarrow s.$$

Tautologies are logically necessary truths. If we can say

$$\vdash Gks \rightarrow s,$$

we can also say

$$\Box(Gks \rightarrow s).$$

Formally in modal logic we use the Rule of Necessitation,[11] which entitles us to infer $\vdash \Box(Gks \rightarrow s)$ from $\vdash Gks \rightarrow s$. We thus have good reason to suppose that if something is known, it necessarily is true.

This necessity is a logical necessity. It arises by the Rule of Necessitation from a tautology. In order to emphasize the distinction between logical and other sorts of necessity, let us for the remainder of this chapter, use capital **L** for logical necessity and the box, $\Box$, for the others. We thus have established, besides the *necessitas consequentis*, $Gks \rightarrow \Box s$ of the last section, a *necessitas consequentiae*

$$\mathbf{L}(Gks \rightarrow s)$$

---

[10] The symbol $\vdash$ represents the 'turnstile' entailment sign.

[11] See below ch. 5, §(i), p. 82.

But such a premiss is not enough by itself to yield worrisome conclusions. In order to derive Ls from L(*Gks* → *s*) we need L*Gks*; but all we have is *Gks*,[12] and all that we can validly infer from those premisses is simply s, that I shall go to Professor Strawson's lecture, not that I must. Nor are we likely to be able to establish L*Gks*: no contradiction is involved in denying that I shall go, or in denying that God knows I shall. An argument from God's omniscience might be attempted, but once again it must fail. It could only show that IF *p* is true, then necessarily God knows it, that is, that it is necessary that if *p* is true then God knows that *p*. It cannot show that God must know that *p* *simpliciter*, because *p* might be false, and God cannot know false propositions.

But it is to misconstrue the argument to take the necessity of God's knowing what I shall do as a logical necessity.[13] Much the most plausible rendering of the necessity operator, □, is that it represents a temporal necessity (which the Schoolmen, rather misleadingly, called a 'contingent' necessity). If God knew yesterday that I should go to Strawson's lecture tomorrow, then by now it is unalterably the case that God knew it yesterday, and in that sense necessary. We are thus entitled to assert

$$\Box Gks,$$

the second premiss we needed in the inference. But what about the first premiss? Are we entitled to assert, in this sense of the necessity operator,

$$\Box(Gks \rightarrow s)?$$

Surely not. Our arguments depended solely on the meaning of the word 'know', which gives rise to a logical necessity, not, so far as we know, a temporal one. The form of the inference then appears to be

[12] See below ch. 5, §(i), p. 84.
[13] See Anthony Kenny, 'Divine Foreknowledge and Human Freedom', in Anthony Kenny, ed., *Aquinas: a Collection of Critical Essays*, London, 1969, pp. 265–6.

$$L(Gks \rightarrow s)$$
$$\Box Gks$$

$$\Box s,$$

which involves mixed modalities, and is not obviously valid as a derivation in modal logic. That, indeed, is true. But the necessity of the whole consequence is a logical, analytical one, and it seems intuitively convincing that an analytic necessity will carry others with it. Jonathan Edwards argues that 'those things which are indissolubly connected with other things that are necessary are themselves necessary': for 'To say otherwise would be a contradiction; it would be in effect to say, that the connection was indissoluble, and yet was not so, but might be broken.'[14] The argument is convincing. If it is temporally possible that I shall not go to Strawson's lecture tomorrow, when it is temporally necessary that God knew this morning that I should go to Strawson's lecture tomorrow, then it is temporally possible that God knows s and yet not-s, in which case it is logically possible – i.e. not inconsistent – to say that God knew s and yet not-s, and hence not logically necessary that if God knew s then s. There is, thus, no fallacy of modal logic in the argument. Provided we have some concept of temporal modality, and provided our analysis of the concept of knowledge justified our taking past knowledge as a hard, unalterable fact, we should have a cogent argument for the unalterability of foreknown truths about the future.

(v) Defeasibility

Two different necessities are involved in foreknowledge. There is some sort of rational necessity involved in the warrant we need to have if we are to claim to have knowledge and not mere

---

[14] Edwards, *Freedom of the Will*, part II, section XII, p. 116, in 1969 edition, ed. Kaufman and Frankena; pp. 165–6 in 3rd ed London, 1760; p. 171, London, 1831; p. 168, Liverpool, 1877.

belief: and there is some sort of temporal necessity involved in
the fact that claims already made in the past entail propositions
which have yet to come true or false. If only one necessity had
been involved, it would have been much easier to unpick the
argument, and see how the mundane fact that we often
foreknow the future could be reconciled with the firm belief we
have that we often can change our minds about what we are
going to do.

Knowledge is two-faced. To borrow a useful distinction from
Toulmin,[15] a claim to know can be assessed both on the score of
whether it was properly made or not, and on that of whether it
was vindicated in the event or turned out to be mistaken. A
claim to knowledge can turn out to be mistaken without having
been improper: and we may stigmatize someone as irresponsi-
ble if he claimed to know something when in fact he did not,
even though his assertion turned out true. Both conditions are
necessary, and if either fails the claim to knowledge has to be
withdrawn, and be redescribed, in the one case as reasonable,
but as it happens, false belief, and in the other as unjustified, but
as it happens, correct belief. In the former case the withdrawal
is retrospective, and even though my claim was responsibly and
properly made, I have to eat my words, and say 'I thought I
knew, but I did not'. It thus appears that future events, and in
particular my future actions, have the power to alter what was
the case so far as your knowledge about the future course of
history, and in particular my future actions, is concerned. It is
like some other cases, which have come to be called 'Cambridge
changes', where I can apparently alter the past. I can make it
true that Julius Caesar died two thousand and thirty years
before I crossed the Rubicon – by taking care to cross the
Rubicon myself exactly two thousand and thirty years after his
death – and I can make it false that Julius Caesar died two
thousand and thirty years before I crossed the Rubicon – by
taking care not to cross the Rubicon myself exactly two
thousand and thirty years after his death. Or again, an employer

[15] S. E. Toulmin, *The Uses of Argument*, Cambridge, 1958, pp. 57ff.

can, by sacking an employee first thing on Monday morning, thereby make the previous Friday the last day the employee worked for him. On Friday it was not true that that was the last day of the employee's employment, but it was retrospectively made true by the employer's subsequent action.[16] In much the same way I can make it true, or false, that you knew that I shall go to Strawson's lecture tomorrow. Whether or not you knew it depends in part on what I shall do. Although 'knew' is grammatically in the past tense, logically it is not entirely past, but has some future reference too, though being none the less definitely dated in the past. In our ordinary way of speaking you already knew that I should go to Strawson's lecture, if you had good reason to expect me to, *even before* the time when I could have made it false. If you had adequate justification for expecting me to go to Strawson's lecture, you could have properly said that yesterday you knew that I would. You knew it unless and until you were proved wrong in the event. With ordinary, non-philosophical foreknowledge of the future free actions of free agents, the agents play the part of the employer, the foreknowers the part of the employee. The agent can Cambridgely bring it about that the foreknowledge of his actions was indeed foreknowledge or that it was not, although the time when it was, or was not, foreknowledge, is indubitably in the past. But since ascriptions of knowledge are defeasible, they do not constitute hard, unalterable facts, and can be revised and re-assessed with hindsight after the event.

Cambridge change seems fishy. It does not enter into the ordinary nexus of cause and effect. A Cambridge change does not produce any other changes in its immediate spatiotemporal vicinity.[17] It has, on those grounds, been stigmatized as not being really real. But that is to claim too much. True, Cambridge changes are not *physical*. But we have no warrant for confining reality to physical reality. Legal and moral analogies may help. In some circumstances one person can enter

[16] I owe this example to Mr C. A. Kirwan, Fellow of Exeter College, Oxford.
[17] H. D. Mellor, *Real Time*, Cambridge, 1981, pp. 107–10.

46   *Foreknowledge and Freedom*

into a contract for another. A father can sign a lease for a grown-up son who is abroad, subject to the son's not repudiating it on his return. If the son repudiates it, then the contract is null and void. But if the son does not repudiate it, the contract is binding, and runs from the date when it was signed by the father, not from when the son implicitly approved it. The father's signing creates legal and moral obligations from that date, but only defeasibly. They can be subsequently annulled by the son's repudiating the contract. But unless annulled, they already exist, and are not created by the son's implicit approval of his father's action on his behalf.

The possibility that we can alter the past in respect of the way it ought to be described and evaluated is no new discovery. Aristotle in his discussion of responsibility makes disclaimers of responsibility conditional on μεταμέλεια (*metameleia*), regret; whether I did not really do something depends not only on the conditions obtaining at the time (that I acted under duress, βία (*bia*) or in ignorance, δι'ἄγνοιαν (*di'agnoian*)), but on subsequent attitude as well; if I am not sorry afterwards that my javelin hit a particular bystander, then I wounded him, even though I had not intended to at the time I threw it. Two less happy examples, one from the Middle Ages and one from a contemporary dilemma, illustrate the same point. Grosseteste, Bishop of Lincoln in the thirteenth century, had a dispute with King Henry II over the legitimisation of illegitimate children. The king's courts held that a child born out of wedlock was a child born out of wedlock, and no subsequent action of the parents could alter that fact. The bishop, believing that marriage was a matter of intention more than of legal formality, held that the subsequent marriage of the parents, provided that they were free to marry at the time of the child's birth, was effective in making the child legitimate all along.[18] In recent

[18] For one of Grosseteste's examples of our being able to alter the truth of statements about the past, see ch. 6, §(i), pp. 105–6. For an account of Grosseteste's quarrel with the king's courts about subsequent legitimisation and some other examples of the past being altered by subsequent events, see R. W. Southern, *Robert Grosseteste*, Oxford, 1986, ch. 10, III, 2, pp. 252–7.

years we have been much exercised over brain death. In many tragic cases we are being led to conclude that a person is dead when the brain is dead, even if the heart is kept beating still by artificial means. But, of course, one cannot be absolutely sure. Miracles can happen, and men in deep coma may yet recover consciousness. We are being led, therefore, to say that the man is dead from the time that his brain stopped functioning, not from the time that the life-support machine was switched off, but that IF a man did recover consciousness, then clearly he was not dead. The ascription of death is defeasible, and should be retrospectively withdrawn in the happy event of subsequent recovery.[19]

Knowledge of the future free actions of a free agent is like a defeasible contract, illegitimacy, or death. That is to say, knowledge is not just a 'hard fact', but contains an element of evaluation, and so is subject to re-evaluation in the light of further evidence. If proved true in the event, it existed all along, from the time the warranted knowledge claim was first made: if, however, it proves false, then it never was knowledge, only justified but untrue belief. In ascribing knowledge, we are not merely stating something physical, but also evaluating the claim to know. There is no physical difference in the immediate vicinity between the case where someone knows what another will do and where he reasonably, but as it happens falsely, opines what he will do. The difference lies in the evaluation of the knowledge claim: in the one case it is vindicated by events, and in the other it is not and must therefore be withdrawn or retrospectively disallowed. Not only with employers sacking unsuspecting employees, but quite generally with the free actions of a free agent that someone else is predicting, the agent has it in his power to bring it about either that the knowledge of his actions was indeed foreknowledge, or that it was not. You know that I shall not let you down. But I still can let you down and make a fool of you all. And equally with God, we have it in our power to turn His foreknowledge into false belief, not

---

[19] I owe this example to M. J. Lockwood, of Exeter College.

because of some failure on His part, but because He has chosen to create a universe in which there are free, autonomous agents, capable of making up their minds for themselves, and each therefore able to change his mind at the last moment and confound even the most reliable predictions about his future action.

(vi) Foreknowledge and Fallibility

Though both philosophical and ordinary knowledge can be ascribed to men, it has traditionally been felt that it would be wrong to ascribe to God mere ordinary, defeasible, fallible knowledge. It would be the most insane impiety, says St Augustine, to say that anything can happen otherwise than God has foreknown it.[20] God is infallible, and His infallibility is what rules out the possibility of our subsequently changing our mind once He has come to know what we are going to do.

But infallibility is a two-edged sword. It cuts out not only human freedom, but, a realm of knowledge. If we stipulate that it is conceptually impossible for God to be contaminated by fallibility, then we are saying that it is conceptually impossible for God to have ordinary, as opposed to philosophical, knowledge. God cannot know where I am going to have lunch, because the only sort of knowledge God can have is infallible, philosophical knowledge, and there can be no such knowledge of future contingents. By making more stringent the standards for divine knowledge, we restrict its range, and make God less, not more, knowledgeable. It is much the same as with power. If we insist that everything God tries to bring about must surely happen, we restrict the operation of the Holy Spirit, which works through human agents, and depends essentially on human cooperation. Many designated Jeremiahs have declined

---

[20] St. Augustine, *De Libero Arbitrio*, III, ii, 4: 'Quisquis enim dixerit aliter evenire posse aliquod quam Deus ante praescivit, praescientiam Dei destruere insanissima impietate molitur.' cf. *City of God*, bk V, ch. ix.

the office of speaking out the truth, because they were too young, too busy, too frightened, or had not been to the right school. Less spectacularly, we have all on occasion failed to do God's will. It is right on such occasions to say that God's will has not been accomplished as originally intended. To make out that since it was not accomplished, it cannot have been intended, is to deny that God could ever work through the willing cooperation of human agents, and thus once again to confine the power of the Deity to DIY jobs. If God, or any one of us, is to be able to work through the agency of other autonomous agents, it is inherent in the whole endeavour that the others may refuse to play, and He, or we, will be unsuccessful. But to say that we can never undertake anything unless we can be sure of succeeding is to diminish our range of power, not to enhance it. And the same holds for knowledge too.

Some philosophers, Nelson Pike among them,[21] have suggested that, to avoid St Augustine's anathema, we do not need to say that God is infallible, but only that He is never actually mistaken in point of fact; and the latter is a different, non-modal notion which carries no modal threat to freedom. From the premises that God believes that I shall go to Professor Strawson's lecture tomorrow and that it is not the case that God is mistaken, all that follows is that I shall go to Professor Strawson's lecture tomorrow, not that I necessarily shall. Thus it might be possible to have a fallible God who is never actually wrong, and this would be compatible with freedom. But that manoeuvre will not work, as the argument of the angels shows.

Angels are unfashionable. Even those who believe in God are embarrassed by angels, and it remains a standing reproach against the Schoolmen that they used to debate how many angels could dance on the point of a needle. Here, however, we are concerned not with the location or looks of angels, but their logical excellences. As their etymology suggests, angels are

---

[21] Nelson Pike, *God and Timelessness*, London, 1970, ch. 4, III, (b), pp. 76–82.

bearers of messages, that is to say in the terminology of modern logic, sequences of propositions. When I talk of a company of angels, I am referring to a set of sequences of propositions, or, more fashionably, to a set of possible world histories.[22]

Suppose, then, there are angels. Suppose there are lots of angels, alephs and alephs of them. Each angel has a different opinion about the course of history, and for every possible course of history from the dawn of creation, there is an angel who opines that that course of history is the actual one. With the passage of time many angelic opinions are proved false – infinitely many. But there remain infinitely many angels whose opinions have not yet been proved false, and are, so far, true. At any future date this will still be the case. Even when the sun turns into a red giant, and finishes off the last human being, there will be some angels whose opinions have all along been true. But no determinist consequences follow. From the fact that some particular angel, Gabriel say, has, as it happens, held opinions all of which have turned out true, it does not follow that any opinion he happened to hold had to turn out true. Gabriel might have thought that I was going to go to Strawson's lecture tomorrow, but if I change my mind and decide not to after all, it simply means that Gabriel is no longer one of the angels whose opinions have always turned out true. Instead of the company around Gabriel, there would have been another company of rightly opining angels, with perhaps Raphael among them. The fact that at the end of history there must be some company of rightly believing angels does not entail that there existed the day before yesterday an already specified company of angels that had to be as of then rightly believing. All that there was the day before yesterday was a company of angels, numbering among them both Gabriel and Raphael, all of whose opinions so far had turned out correct. Whatever I decide to do tomorrow, there will be some angels who will turn out to have opined correctly. But the company of rightly believing angels is not determinate. There is not a determinate

---

[22] In the terminology of ch. 9, a branch.

company of rightly believing angels now, nor has there ever been one: rather, the company of rightly believing angels was determined yesterday by my deciding not to go off to Paris instead of keeping my engagements at home, and will be determined tomorrow either by my going to Strawson's lecture or by my not going. Thus the existence of a company of angels who at the last day turn out to have had true opinions about every putatively free choice of every human being does not pose a threat to freedom, because their opinions, although as it happened true, were not necessarily true. At the last day, Gabriel, or Michael, or whoever is the spokesman for the rightly believing angels, will say with proper modesty 'We are not infallible: it is just that we were never wrong.'

It seems reasonable to consider God among His angels, and to ascribe to Him an opinion about the course of history, and to hazard the guess that at the last day God will be numbered in the company of rightly believing celestial beings. In that case, it would seem, God would turn out to have been omniscient, without having been incapable of being wrong. It was not in the least impossible for God to have been wrong any more than it was impossible for Gabriel or Raphael to have been wrong. It just so happened that God turned out to be right. After all, someone has got to be right: why not God? In this way we might hope to preserve human freedom by imputing to God not a necessary omniscience, but only a contingent one. God is a better bet than any of the angels who disagree with Him, because although it is perfectly possible for me or you or Adam to prove God wrong and Raphael, say, right, it is also possible that none of us will and that He will turn out to be right, and He knows us better than Raphael does.

But there is an important difference between God and the rightly believing angels. The company of rightly believing angels is indeterminate until the last day, but God is determinate from the very first. The fact that there exist rightly believing angels poses no threat to freedom because there also exist wrongly believing angels, and only time will tell, in the light of our free choices, who the rightly and who the wrongly believing angels are. But there is only one God, and His opinions about the course of history represent a determinate set of opinions which can be specified now, in advance of the last day, and not left until the last day shows which of them can be picked out as being *ex post facto* true. If I think some angel has a true opinion about what I shall decide to do tomorrow, I am not jeopardising my freedom of decision, because as of now there are angels opining either way, and tomorrow will show which of them opined rightly. I am not saying that there are some angels I could conceivably name – *sunt quidam angeli* in Latin – who rightly opine what I shall do, but only that the company of rightly believing angels will turn out not to be empty. In the case of God, however, He is a definite being and His opinions are already determinate. Tomorrow can tell whether they are right or wrong, but not what they are or who He happens to be. To make God's omniscience purely contingent and *ex post facto* would be to deny His unique status. He is not merely one among innumerable angels who happens to have been right all along, but is, if He is anything, the one true God, entirely different from all other beings. He is a specific individual, and if He has opinions about the future actions of free agents, He has them. They are, so to speak, nailed to the mast of His uniqueness. If God yesterday believed that I shall go to Strawson's lecture tomorrow, then it is now unalterable that God has this opinion. If, furthermore, all His opinions are true, not just in the valedictory *ex post facto* sense[23] but as and when He holds them, then it is now unalterable that God has the true opinion that I shall go to Strawson's lecture tomorrow; in which case it is now unalterably true that I shall go to Strawson's lecture tomorrow. Angels may make guesses, some of which may turn out to be true, and some angels will prove to have guessed correctly all along: but God, because He is God, cannot just be guessing. If He has opinions, we are concerned not only with their valedictory *ex post facto* truth, but with whether they were true at the time they were held.

[23] See below, ch. 4, §(iii), pp. 65–72.

And hence it does not seem possible that God should be merely contingently omniscient, although not incapable of being mistaken.

There is a parallel between the argument of the angels and various arguments for logical determinism. There is an essential difference in the order of quantifiers and the status of the entities being quantified over, which is difficult to articulate and difficult to disentangle. These will be discussed in chapters 7 and 9.[24] So far as the present argument is concerned, it is evident that there is no escape from St Augustine's objection by hypothesizing a fallible but not ever actually mistaken God. Forebelief poses a problem for divine perfection to which we shall return in chapter 11.

(vii) Foreknowledge and Foretruth

Knowledge involves not only a knower but something known, and it is easy to objectify what is, or might be, foreknown so as to make that into a fetter on human freedom. Kenny objects to a suggestion of mine that God might have voluntarily forgone foreknowledge of our future actions in order to preserve our freedom of choice on the grounds that 'the knowledge renounced by God must be knowledge which it is logically possible to have: hence Lucas's solution does nothing to solve the problem of how divine foreknowledge can be logically reconciled with future indeterminism.'[25] Kenny's picture is one of there being knowledge there for God to know, but of His turning a blind eye to it; God so much respects our privacy of intention that instead of knowing all there is to know, He chooses not to know some of the things there are to know, much as a father does not look into his son's room before Christmas so that his son's surprise present really shall be a surprise. Certainly if that were the case we should be committed

---

[24] Ch. 7, §(ii), pp. 125–8, and ch. 9, §(iii), pp. 166–70.
[25] Anthony Kenny, *The God of the Philosophers*, Oxford, 1979, pp. 60–1.

to fatalism. If the future were in that sense knowable, then it is fixed and unalterable: τὸ παντελῶς γνωστὸν παντελῶς ὄν (*to pantelos gnoston pantelos on*), what is altogether knowable is altogether real, because the knowable is there waiting to be known, whether actually known or not. All God does, as Kenny understands my claim, is not to pry into matters it would embarrass us for Him to know ahead of time. But this is not the suggestion I had in mind. It is not that God has chosen to avert His eyes, and not to know things He could perfectly well know, but rather that having chosen to create a world in which men are free, the things that would have been knowable, had He created a different, deterministic world, are not there to be known at all.

Traditional philosophy finds it difficult not to objectify knowledge, and suppose that there are objects of knowledge which, whether actually known or not, have some independent existence. But if ordinary knowledge is defeasible, then there is no substantial knowledge there to be known independently of what a knower can legitimately be said to know. And this has implications for the account we must give of truth. Only if there is no independent, constraining truth or falsity to propositions, is it reasonable to say that God's omniscience is not compromised by His not knowing them. Essentially, I am denying 'hard', 'indefectible' truth to future contingents. If I am to put forward a defeasible theory of ordinary knowledge, I must put forward also a defeasible theory of truth. To this task I now turn.

4

Truth and Bearers of Truth

(i) The Many Faces of Truth

Even at its most pedestrian, truth is tricky. The word 'true' is used in part as an operator, to assert and warrant, to endorse and commend, to concede, to entertain, and to suppose.[1] It is used also as an adjective, to describe and evaluate utterances, sentences, statements, predictions and the like; and in the mouths of logicians it has become a platonist term of art, ascribing a metalogical truth-value to abstract entities in the propositional calculus. Philosophers have tended not to distinguish these different uses. They think about truth, and carefully stipulate how the word 'true' is to be used, and then in their unthinking moments relapse into using it in other ways. Words are almost as conservative as people. You can make a word learn new tricks, but seldom forget its old ones; and 'true' carries with it into its respectable middle age among logical constants and well-formed formulae the habits of an operator it learned in its youth.

The word 'true' is used to commend what is said as worthy of belief. Etymologically, it is connected with the German *treu* – meaning faithful or reliable, a use which continues in the phrases 'twelve good men and true', 'a true friend', 'a true die' –

---

[1] R. M. Gale, *The Language of Time*, London, 1968, pp. 138ff.; P. F. Strawson, 'Truth', *Analysis*, 9, 1948–9, pp. 83–97. See also J. L. Austin 'Truth', *Proceedings of the Aristotelian Society*, supp. vol. 1950, pp. 111–28; W. H. Walsh, 'A Note on Truth', *Mind*, 61, 1952, pp. 72–4; A. R. White, *Truth*, London, 1970, ch. 3 (a), pp. 41–56.

and with the English words 'trust' and 'troth'. To say that an utterance is true is to assure the hearer that he can trust it, that it is reliable, and will not let him down if he pins his faith on it. It is like the word 'probable'; only, whereas the word 'probable' gives only guarded guidance, the word 'true' gives an unguarded guarantee. Instead of hedging, as with the word 'probable' or the parenthetical phrases 'I think' or 'I believe',[2] the word 'true' is more like 'I know' or 'I promise that', and gives an unconditional warranty.

The warranty determines the subsequent development of the concept. If I claim that it is probable that a penny will come down heads at least once in the next ten tosses, you cannot hold me responsible if in fact it does not. My claim, to use again Toulmin's distinction,[3] was a perfectly *proper* one, even though it turned out to be mistaken, and did not, in fact, come true. Provided I had good reason for making the claim, you have no recourse to me if it is subsequently falsified by events, because I had specifically disclaimed responsibility. With the word 'true', however, it is different. Although Toulmin's distinction between proper and mistaken claims remains, and you could not chide me for having deceived you or misled you if my claim that it was true that Paul would be at the party, properly made on the basis of what he himself had told me, turned out to be incorrect, I should none the less have to withdraw my original claim, and, instead of saying 'Well, it was true, even though he did not in fact turn up', say rather, 'Well, I thought it was true, even though it turned out not be be'. Because the word 'true' gives an unconditional guarantee, it has to be retrospectively withdrawn, should it turn out to have been mistakenly used, no matter how proper its use was at the time of the original utterance.

The word 'true' in this usage is like the word 'know'. Both are defeasible. But defeasibility is two-sided. Much as knowledge has been taken to be infallible because it would not have been

---

[2] J. O. Urmson, 'Parenthetical Verbs', *Mind*, 1952, pp. 480–96.
[3] See above, ch. 3, §(v), p. 44, n.15.

knowledge if it had turned out to be false, so truth, defeasible as a claim, is understood as being 'indefectible' in content. 'Call no man happy until he is dead', said the Greeks: to be properly happy, the hero and heroine must live happily *ever after*. Since a claim to truth must be withdrawn, no matter how responsibly and properly made, should it subsequently turn out to be false, it follows that *if* it is to be really true, there must be no possibility of its subsequently proving false. Truth is taken to be indefectible because the unconditional warranty, once issued, applies all ways, and so for all subsequent time. Indefectible truth has a stability that probability lacks, and whereas probabilities and some other sorts of truth can change, indefectible truth cannot. Indefectible truth is governed by the principle 'once true, always true'. It is a temporal sort of truth. We can ask whether something was, or is, or will be indefectibly true at a certain time, and although if it is indefectibly true at any one time, it must be indefectibly true at all subsequent times, it does not follow that it must have been true at all earlier times. At some earlier times it may not have yet come true. Indefectible truth can be acquired, but not lost.[4]

It is temporal, but not omnitemporal, and because we are free to take up any temporal standpoint,[5] we are impelled to go further, and develop an omnitemporal concept of truth, to enable us to assess at different times propositions expressed by utterances at different times. In this sense I can say that what you said yesterday *was* true, denying your critic's claim who said yesterday 'It isn't true.' Although it is clearly a temporal use of the word 'true' – we use tensed verbs with it – there is no possibility, in this sense, of something's *coming* true: either what you said was true or it was not.[6] We are concerned solely

---

[4] C. D. Broad, *Scientific Thought*, London, 1923, ch. 2, p. 82.

[5] See above, ch. 1, pp. 11–12.

[6] cf. Cicero, *De Fato*, 12, 28: 'nec, si omne enuntiatum aut verum aut falsum est, sequitur ilico esse causas immutabilis easque aeternas, quae prohibeant quicquam secus cadere atque casurum sit: fortuitae sunt causae, quae efficiant, ut vere dicantur quae ita dicentur "veniet in senatum Cato," non inclusae in rerum natura atque mundo. et tamen tam est immutabile venturum, eum est

with the invariant core of what was expressed by the utterance, not with whether the utterance was responsibly made at the time of utterance. The word 'true' is being used not as an assertion operator, warranting the assertions we actually make, but as an evaluative term which we can use for an all-time assessment of propositions. We are giving a verdict, rather than vouching for evidence.

We do not only judge. We wonder, we hypothesize, we argue. We wonder whether a hypothesis is true, and suppose it is for the sake of argument. 'I don't know', a juror may say in the jury box, 'whether what the witness said was true.' 'Suppose, for the sake of argument, it was,' says another; 'then why did he tell a different story to the police?' In these uses the word, although clearly used in tensed discourse, is not earthbound by the burden of responsible utterance, but is a wide-ranging logical counter that can float freely on currents of speculative thought, untrammelled by ties of time or temporal context, concerned solely to distinguish the true propositional sheep from the false propositional goats.

The use of the word 'true' as a logical counter is independent of time, and in any case the temporal uses of the word 'true' are difficult to make sense of logically, and logicians, moved partly by platonist urges, partly by the desire to disengage themselves from problems of temporality, have preferred to use the word in a way which is not merely omnitemporal but so much transcends time as to be absolutely timeless. 'True' then comes to designate a paradigm platonist property, which is predicated tenselessly, and will be distinguished hereafter by a capital letter[7] – True. In the same way the word 'False' will be

---

verum, quam venisse, nec ob eam causam fatum aut necessitas extimescenda est. etenim erit confiteri necesse "si hoc enuntiatum 'veniet in Tusculanum Hortensius' verum non est, sequitur falsum sit." quorum isti neutrum volunt, quod fieri non potest.' and 16, 37: 'sed ex aeternitate vera fuit haec enuntiatio "relinquetur in insula Philoctetes," nec hoc ex vero in falsum poterat convertere'.

[7] Except in ch. 8, §(iii), p. 146.

predicated tenselessly, so that we can say that some propositions, or propositional contents, or Quinean eternal sentences, *be* True or *be* False.[8] The distinction between omnitemporal truth and timeless truth is difficult to draw. It seems reasonable to regard the laws of nature as omnitemporally true, and theorems of mathematics as timelessly true, but the arguments are not conclusive, and for our purposes it is not necessary to decide the issue. In mathematical logic, in fact, although we have occasion to use meta-logical properties, and discuss whether some proposition *be* True or *be* False, we operate mostly with two truth-values, which are not properties but abstract objects, mathematically characterized as the members of the Boolean ring, B_2. I shall refer to them by the words TRUE and FALSE in capitals; most commonly the letters T and F are used; sometimes $\top$ and $\bot$; sometimes 1 and 0. The difference between 'TRUE' and 'True' is that the former is a noun referring to an abstract entity, the latter an adjective ascribing a platonist property. The two uses are linked. To say that a proposition *be* True is the same as to say that it *have* the truth-value TRUE.

The mathematical logicians' concept represents the extreme development in the direction of the unconditional guarantee implicit in the ordinary use of the word 'true'. The ascription of the property True and the assignment of the truth-value TRUE has nothing to do with responsible assertion, and everything to do with non-mistakenness. It simply duplicates at the metalogical level whatever there is in the proposition itself at the ordinary, logical level: as Tarski has defined it, '*p*' be True if and only if *p*.[9] The definition seems a trivial truism, but is of fundamental importance. It suggests both the dispensability and the indefinability of the concept of truth, and has led philosophers in both directions.

---

[8] For the use of the italicised infinitive part of the verb to express tenselessness see ch. 2, §(i), p. 16, n. 4.

[9] Alfred Tarski, 'The Semantic Concept of Truth', *Philosophy and Phenomenological Research*, 1943–4, pp. 341–75; reprinted in H. Feigl and W. S. Sellars, *Readings in Philosophical Analysis*, New York, 1949, pp. 52–84.

Tarski's definition seems open to Ramsey's complaint that it is vacuous in content – economy would suggest a 'No Truth' theory.[10] Instead of talking pompously in the metalanguage about propositions *being* True, we can equally well, and more simply, assert those same propositions in the object language. But logicians do not use the word 'True' only to state in the metalanguage what could equally well be stated in the object language without it. They use it to talk about and understand the object language, and in particular to make models. A semantic model of a formal system is an interpretation of its terms which brings out the axioms as True, and the rules of inference as truth-preserving. More formally, an interpretation is given by means of a 'truth-valuation function', that is to say a function from the well-formed formulae of the formal system onto the two-element Boolean ring, B_2, whose members are TRUE and FALSE. In chapters 8 and 9 we shall be developing a semantic model for tense logic which not only helps give an intuitive understanding of the formalism but validates the axioms and rules of inference we had proposed, by showing that they do, indeed, hold in all interpretations of the relevant kind.

Tarski's metalogical definition of truth suggests the possibility of devising an entirely logical definition of the concept. But this, he proved, is impossible. Tarski's theorem is one of a cluster of incompleteness theorems, of which Gödel's theorem is the best known, which have entirely changed the aspect of philosophy in the twentieth century. It shows that we shall never be able to give a satisfactory definition of truth in a completely formalised logical system that is adequate for elementary arithmetic. For if we are to have any meta-logical account of truth at all, it must satisfy Tarski's definition, and if it were expressible in a formal logical system, it could be coded by some Gödel numbering procedure, which in turn could be represented in the logical system itself, provided that it was adequate for elementary number theory. And then we could

---

[10] F. P. Ramsey, *The Foundations of Mathematics*, Cambridge, 1931, pp. 141–2. See further below, ch. 8, §(iii), p. 148.

construct a formal version of Epimenides' Cretan Liar paradox, that this statement is itself untrue, which would not just be a paradox, but a straight self-contradiction. The only way to avoid the contradiction is to reject the assumption that truth can tied down by some neat and unambiguous definition. Truth, we are forced to conclude, is an irreducibly intensional concept, which can never be adequately represented in entirely external terms, but needs to be grasped and understood by a mind.[11]

There is an intimation of the intensional and personal character of truth in its traditional identification with God, which leads St Anselm to argue from the omnitemporality of truth to the eternity of the Ultimate Reality.[12] Anselm's argument is weakened by there being many different sorts of truth, but the identification expresses important insights, showing not only the ineliminable intensionality of truth, but its ineffable indispensability. It is impossible to tie down truth by neat unambiguous definitions, but it is none the less a concept we possess, knowing within ourselves the pull it exerts on our searching and believing. We want to know the truth, and reckon that only true statements are worthy to be believed. For many scientists and academics the truth is the only worship they know, and many others have come to God as the Truth, before knowing Him as the Way or the Life. Often, indeed, there has seemed to be an incompatibility between the cold objectivity of the truth and the personal commitment that the worship of a personal God requires; and many have felt obliged, in the name of truth, to deny that the ultimate reality be personal. But if the ultimate reality be not personal, then personal categories of thought have no ultimate grounding in reality, no ultimate

---

[11] Dale Jacquette, 'Metamathematical Criteria for Minds and Machines,' *Erkenntnis*, 27, 1987, pp. 1–17, esp. §7.

[12] St Anselm, *Monologion*, ch. 18: 'quare idem sequitur de summa natura, quia ipsa summa veritas est'; in J-P. Migne, *Patrologiae*, 158, Paris, 1863, p. 168; tr. J. Hopkins and H. Richardson, *Anselm of Canterbury*, London, 1974, vol. I, p. 29; quoted by A. J. Moen, *God, Time and the Limits of Omniscience*, Oxford D.Phil. Thesis, 1979, ch. 1, p. 18.

purchase on us. And among those personal categories of thought we have to reckon truth, which commends itself only to free, autonomous minds.[13] Only if the ultimate reality is personal can we be under an obligation to seek the truth and believe it. So that those who claim that reality is something less than personal, and seek to reduce all personal categories of thought to something simple and more mechanical, belie the content of their claim by their very action of commending it to us as true.[14]

Tarski's theorem greatly alters our understanding of the concept of truth. It is an intensional concept that can be grasped only by a mind, so that we cannot hope to eliminate its Protean proclivities by rational reconstruction, which could then be applied according to some rigidly defined, algorithmic criterion. We may portray its many faces, but cannot replace them by a single blue-print.

(ii) Bearers of Truth

Because there are different senses of the word 'true', there are different sorts of things that can be said to be true, different bearers of truth, as I shall call them. Each different sort of truth impresses on its bearers a different logical shape, and conversely, as we consider different sorts of bearers, we devise different sorts of truth.

When we are using the word 'true' as an operator, to assert and warrant, to endorse and commend, or to concede, it is

[13] See Warner Wick, 'Truth's Debt to Freedom', *Mind*, 82, 1964, pp. 527ff.; E. L. Mascall, *Christian Theology and Natural Science*, London, 1956, ch. 6, 2, pp. 212–219; J. D. Mabbott, *Introduction to Ethics*, London, 1966, pp. 115–16; J. R. Lucas, *The Freedom of the Will*, Oxford, 1970, §21, pp. 114–16, and the other authors cited on p. 116.

[14] Compare J. E. McTaggart, *Philosophical Studies*, London, 1934, p. 193: 'If materialism is true, all our thoughts are produced by purely material antecedents. These are quite blind, and are just as likely to produce falsehood as truth. We thus have no reason for believing any of our conclusions – including the truth of materialism, which is therefore a self-contradictory hypothesis.'

utterances,[15] or 'token-sentences' as the Kneales call them,[16] that it is operating on. If, however, I am interested not only in the words 'It is raining' or 'Socrates is sitting' in your mouth on a particular occasion, but in anyone's mouth on many occasions I shall ascribe truth to sentences – sentence-types, as we might call them for the sake of clarity – as well as to sentence-tokens. Often it makes a difference when or by whom the sentence was uttered. 'It is raining' said by me now is, as it happens, false, though if you were to say it in Manchester, it might well be true, as also if I or anyone else were to say it on many other occasions in Cambridge. 'Socrates is sitting', if the name 'Socrates' is taken to refer to the teacher of Plato, is now false, but two thousand four hundred years ago was sometimes true and sometimes false. If these sentences are the bearers of truth, the sort of truth they bear is a mutable truth, quite different from the indefectible truth we distinguished in the previous section.[17] Many different utterances are utterances of the same sentence, and sentences can be regarded as particular classes of utterances, so it is not surprising that they bear a different sort of truth. We can distinguish other classes of utterance – all those that mean the same in English (e.g. 'Brutus killed Caesar', and 'Caesar was killed by Brutus) – all those that mean the same in other languages as well as English (e.g. 'Brutus killed Caesar' and '*Brutus Caesarem interfecit*') – and we can regard each as a bearer of truth of an appropriate sort.

My actual words soon fade into silence and may well be lost beyond recall, but what I said still stands for assessment as regards truth and falsity, in the same way as it can be denied or affirmed at other times and by other people. Indefectible truth, therefore, and likewise omnitemporal truth and timeless truth demand bearers which are much less fleeting than words can ever be. So we postulate such bearers, and devise names for them. Such names are terms of the logician's art. Different

[15] P. F. Strawson, 'True', *Analysis*, 9, 1948–9, pp. 83–97.

[16] W. C. and M. Kneale, *The Development of Logic*, Oxford, 1962, p. 49.

[17] For another example of mutable truth, due to Grosseteste, see ch. 6, §(i), p. 105.

logicians define them differently, some talking of statements, some of sentences, some of propositions, some of propositional contents, some of well-formed formulae and some of judgements, with importantly different meanings in each case.[18] There is much dispute about what exactly they are. But for our present purpose it will be enough to regard them as abstract entities of some sort, and to call them *propositions*, with a sub-class of *propositional contents*. The term 'proposition' denotes an abstract entity that is the common core of what is expressed by tensed sentences and is invariant under different occasions of utterance. It is what is expressed either by a tenseless sentence-type or by a tensed utterance in which all temporal references have been made explicit, such as 'There will be a sea battle tomorrow, February 29th, 2000', 'There is a sea battle today, February 29th, 2000', 'There was a sea battle yesterday, February 29th, 2000'. It is what may be asserted or denied by the same or different speakers on the same or different occasions of utterance: just as the negation of one on one occasion contradicts the assertion of the corresponding utterance on another occasion, so the omnitemporal truth of the one carries with it the omnitemporal truth of the other.

Propositions typically are of an RE form, and are expressed by different utterances on different occasions with an appropriate S but the same RE. *Propositional contents* are minimal propositions, Reichenbach's E, not viewed from any temporal perspective. They are Quine's eternal sentences, or Rescher's temporally definite propositions.[19] Typical RE propositions which differ only in their reference point R, have the same propositional content, E, but view it from a different temporal perspective. It is reasonable to define proposition so as to include propositional contents. For the Reichenbachian exegesis of the simple future, the present and the aorist posits a reference point R that is not always apparent to users. Reichenbach

[18] W. C. and M. Kneale, *The Development of Logic*, pp. 49–51.
[19] W. V. O. Quine, *Word and Object*, New York, 1960, pp. 193–4, 208, 226–7. Nicholas Rescher, 'Truth and Necessity in Temporal Perspective', in R. M. Gale, ed., *The Philosophy of Time*, London, 1968, pp. 183–220.

analyses the simple future as (S–R,E), but to many it seems to be just (S–E). We can accommodate Reichenbach's scheme by supposing there to be an unnoticed R, but then we are effectively treating a simple E and an RE, with the R contemporaneous with the E, as equivalent; in which case we do not want to make a deep distinction between propositional contents with just an E and the equivalent propositions with a contemporaneous R. It is sensible, therefore, to include propositional contents with propositions, as being both bearers of the same sorts of truth.

It is not clear what sorts of truth can be ascribed to propositions and propositional contents. They are terms of art, and we have few intuitions from everyday use to guide us. There is evidently an argument for describing a proposition as omnitemporally true, and propositional contents can clearly come true in a temporal, indefectible way. Rather than start with these locutions, however, it is best to start with the sort of truth logicians are most ready to ascribe to propositions, namely timeless truth, and from that construct a model of omnitemporal and simple temporal truth. This is done in chapters 6, 8 and 9.

(iii) Conjectures and Predictions

Since there are different sorts of truth and different bearers of truth, we need to walk warily in ascribing truth, and in particular to future contingents. We must always ask 'Future contingent *what?*', and not only distinguish future contingent utterances from future contingent sentences and future contingent propositions and propositional contents of various sorts, each capable of bearing different sorts of truth, but try to distinguish, so far as we can, the different future tenses employed.

Utterances expressed in the future tense need to be distinguished into *predictions* and *conjectures*. A prediction is, logically speaking, couched in Reichenbach's posterior present tense, S,R–E; if it is to be true it needs to be well warranted; it needs to be based on grounds, grounds available to the speaker at the time it was uttered; in the language of the Schoolmen, the future must be already 'present in its causes', so that I am in a position to say then 'It is *going* to happen'. Else, it was, in Toulmin's terminology, improperly made, so far as being a prediction is concerned, and should be regarded as being only a conjecture or a guess. A conjecture can prove to be true, but its truth is of a different sort; it turns entirely on the event; if the event is as was conjectured, the conjecture was true, if not, not. It is like a bet. If I lay a bet with my bookmaker that Eclipse, to use Ryle's example,[20] will win the Derby, and Eclipse in due course does win the Derby, then my conjecture has turned out true, and I collect my winnings. Its truth cannot be impugned on the score of my not having adequate warrant at the time I laid the bet. In saying 'Eclipse will win the Derby', there was no present reference in what I said. I used the future tense because the time of utterance, S, was before the event spoken about, but I was considering it only from a future reference point. Logically speaking, I spoke in the simple future tense, S–R,E, and the appropriate sort of truth is, in Ryle's phrase, 'valedictory truth'.[21] *Ex post facto* we can say that my guess that Eclipse would win came true. It was true after the event, and after the event we are prepared to antedate the ascription of truth and say that it was true before the event. Valedictory truth, though couched in the past tense, is not really past at all, but only, as we shall see,[22] a *façon de parler* for saying the forecast was in fact fulfilled. It conceals a covert future reference and so lacks the unalterability of the genuine past. It can be retrospectively applied after the event, but its retrospective application does not confer a genuine temporal truth at the time the conjecture was made.

Predictions are more substantial, and need a more substantial sort of truth. If they are to be true predictions, they need to be well warranted at the time they are made, and so there must be something at the time they are made in virtue of which they are

---

[20] Gilbert Ryle, *Dilemmas*, Cambridge, 1954, 'It was to Be', p. 22.
[21] *Ibid.*, p. 20.
[22] In ch. 9, §(vii), pp. 176–7.

properly-made predictions, and it is very natural then to identify those truth-making factors with some fact that is already temporally true and necessitates the prediction's subsequently proving true in the event. The strength of this argument can be gauged not only by the appeal that arguments for fatalism have for us, but by the counter-intuitive lengths people are prepared to go to in order not to seem to be committed to already existing future facts. Broad raises the question 'If the future, so long as it is the future, be literally nothing at all, what are we to say of judgements which profess to be about the future?'[23] and concludes that judgements '. . . which profess to be about the future are not genuine judgements when they are made, but merely enjoy a courtesy title by anticipation, like the eldest sons of the higher nobility during the lifetime of their fathers.'[24] The Law Courts have gone even further, and have held that predictions are neither true nor false. In a series of decisions the Court of Appeal held that section 14(1) of the Trades Description Act, 1968, '. . . has no application to statements which amount to a promise with respect to the future, and which therefore at the time when they are made cannot have the character of being either true or false . . .'[25] and that the section is limited to statements of fact, past or present, and does not include assurances about the future.'[26]

A statement that a fact exists now, or that it existed in the past, is either true or false at the time when the statement was made. But that is not the case with a promise or a prediction about the future. A prediction may come true or it may not. A promise to do something in the future may be kept or it may be broken. But neither the prediction nor the promise can be said to have been true or false at the time when it was made.[27]

In a later case the House of Lords reviewed this reasoning and restricted its application, but did not overturn the main argument

[23] Broad, *Scientific Thought*, p. 70.
[24] *Ibid.*, p. 73.
[25] *Beckett* v. *Cohen* [1972] 1 W.L.R. 1596.
[26] *R.* v. *Sunair Holidays Ltd.* [1973] 1 W.L.R. 1113.
[27] *R.* v. *Sunair Holidays Ltd.* [1973] 1 W.L.R. 1113.

that a prediction is not 'a statement as to existing fact which may be true or false'.[28] But the argument is clearly misconceived as regards predictions.[29] We often do describe predictions as true or false: to quote Ryle:

In characterizing somebody's statement, for example a statement in the future tense, as true or false, we usually, though not always, mean to convey rather more than that what was forecast did or did not take place. There is something of a slur in 'false' and something honorific in 'true', some suggestion of the insincerity or sincerity of its author, or some suggestion of his rashness or cautiousness as an investigator.[30]

With predictions we are not happy with simply non-suiting the question of their truth or falsity. Nor does the valedictory sense of 'true' capture what we mean when we characterize them as true at the time of their being made, meaning thereby not only that what was predicted actually came to pass, but that it was warrantedly assertible at the time. In saying that somebody's prediction was a true one we are saying more than that he made a guess which happened to be correct. We are saying that his guidance was trustworthy because it was given by a reliable man who had adequate grounds for warranting his assertion. Although, as we have seen, if his prediction is not borne out in the event, he must, all his grounds notwithstanding, retro-spectively withdraw his truth claim, the bare fact that events turn out as he said is not enough by itself to justify it. It is a necessary, but not a sufficient, condition. What is needed in addition to the actual success of the prediction, are some conditions, obtaining at the time of utterance, which consti-tuted adequate justification. And thus, whereas 'coming true' is an *ex post facto* judgement, depending only on the predicted event's actually taking place, 'being true', as applied to

[28] *British Airways Board* v. *Taylor* [1976] 1 All E R 68; see also pp. 73, 74, 76–8.

[29] A. R. White, 'Trade Descriptions about the Future', *The Law Quarterly Review*, 90, Jan. 1974, pp. 15–20. I am indebted to Professor White for many valuable insights, as well as for drawing these legal judgements to my attention.

[30] Ryle, *Dilemmas*, ch. 2, 'It was to Be', p. 18.

predictive utterances, depends also on the circumstances at the time the prediction was made. If I predict, rather than merely guess, that there will be a sea battle tomorrow, and my prediction is true, then I must have adequate reasons today for saying that there is going to be a sea battle tomorrow – for example, that our admiral has already made up his mind to attack – and so there must be a sea-battle-making factor already in existence in virtue of which it is today the case that there is going to be a sea battle tomorrow. 'True' when applied to predictions is not just a courtesy title conferred on the heir apparent in anticipation of his succeeding to his father's title in due course, but a substantial honour applied in confident expectation of success, but presumptive and subject to retrospective withdrawal, should the prediction fail to be vindicated in the event.

It is easy to agree with the prescription of what we want of predictive truth, difficult to give a coherent account of it. It is uneasily balanced between valedictory truth on the one hand and what Gale calls 'metametalinguistic' truth on the other.[31] 'Metametalinguistic' truth concentrates exclusively on the propriety of the predictive claim. The bearers of metametalinguistic truth are utterances made under conditions of imperfect information, and the condition of their being true is simply that they were adequately supported by the information available at the time of utterance. The grammar of the word 'true' in its metametalinguistic use resembles that of the word 'probable'. We discount altogether the actual mistakenness of a claim, so long as it was reasonable to make it, and would allow 'It was true, although it didn't happen', just as we allow 'It was probable, although it didn't happen'. This use parallels the example of the surgeon who diagnoses cancer, and tells the patient that he is going to die of cancer, although as it happens the patient is so upset that crossing the street on his way back home he is run over by a bus. The doctor might claim that he had given a true prognosis, although it had not proved true in

[31] R. M. Gale, *The Language of Time*, pp. 144–53.

the event. The patient was going-to-die-of-cancer. So too, it would be metametalinguistically true to say we were going to fight a sea battle once the admiral had made up his mind and given the orders, even if an entirely unexpected storm intervened on the morrow and prevented the battle's actually taking place. 'We were going to, but we didn't' would be an entirely acceptable locution in that sense of 'going to'. We are considering the prediction just simply at the time it was made, and evaluating it from the point of that time only, ascribing truth if at that time it was as well based as it could have been. But it is an awkward usage. More naturally we would talk of the the doctor's prognosis being correct, rather than of his prediction being true. We are not comfortable in saying 'It was true when he said it, although it wasn't in the event', and with good reason. The function of predictions is to give guidance, and not to be immune from criticism whatever the subsequent course of events.

There are other concepts with a comparable dual aspect. In our ordinary understanding I cannot be drunk unless I both have imbibed alcoholic liquor and am now intoxicated. If I had satisfied the former condition, but by some quirk of metabolism remained as sober as a judge, I could not be held to be drunk but could only be congratulated on my strong head: equally in our ordinary, non-legal parlance I could not be said to be drunk if I had consumed no alcohol, even though I displayed all the symptoms of being drunk. 'Drunk' is an 'upshot' word, characterizing the present in terms of past events leading up to it. Predictive truth similarly has one foot in the present and the other in the future. Two other examples are offered by White, who compares the word 'true' with the words 'fatal' and 'beginning'. In saying 'The devaluation of the pound was a fatal mistake' or 'was the beginning of the decline of Britain as a world power', we are talking about the devaluation of the pound at the time the pound was actually devalued, although in the light of subsequent events.

The devaluation of the pound was, if indeed it ever was, at the time of its happening the beginning of the decline of Britain or a fatal mistake.

Similarly, the statement about the future was true, if indeed it ever was true, at the time it was made. . . . Hindsight often shows things to have had characteristics which foresight could not have predicted; it does not show them to have later acquired these characteristics.[32]

A prediction is not true *unless* the event subsequently takes place, but that does not mean that it is not true until the event takes place, just as you could not have known I was going to Strawson's lecture unless I did, but that does not mean you did not know it until I did.

Since predictions are intended to give guidance at the time they are uttered and heard, they are susceptible of assessment at that time, and so can be ascribed temporal truth. But equally, since their function is to give guidance, they are intended to rule out certain future possibilities, and are thus vulnerable to refutation if what was said to be going to happen does not in fact happen. They are therefore bearers of temporal truth which is not completely post-dated, as is valedictory truth, but which can be defeated by the subsequent turn of events. It is a temporal truth, defeasible at the time in point of fact, but indefectible in aspiration and in retrospect. We can properly ask whether a prediction is true at the time it was made: if it really was, then it was indefectibly true, and must be adjudged true at all subsequent times; but our claim that it is true is a defeasible claim that may have to be withdrawn in the light of subsequent events.

The defeasible ascription of indefectible truth is modally complex. It only makes sense if the modalities involved are different, and it is difficult to distinguish the possibility of defeat from the impossibility of defection. But once we have done this, we can account for the predictive truth not only of actual utterances but of hypothetical utterances, such as if someone had said ten thousand years beforehand that this would be white,[33] and of the RE propositions that future tensed

---

[32] Alan R. White, *Truth*, London, 1970, pp. 45–6.
[33] Aristotle, *De Interpretatione*, 9, 18^b33–6. But see further, ch. 7, §(iii), p. 129, n. 15.

utterances, or on other occasions other suitably tensed utterances, state. They too can be assessed for truth, and similar arguments will apply. Essentially, where the reference point is contemporaneous with, or later than the event no problem arises, but where it is earlier, as in the posterior present, S,R–E, tense, we find it difficult to ascribe temporal truth.

(iv) Bivalence

There are many pressures on the concept of truth, due in part to the different purposes we need it to fulfill, in part to the different logical shapes of its bearers. One pervasive pressure stems from its use as a logical counter in argument, especially in arguments by dilemma and by *reductio ad absurdum*. In this use the word 'true' serves not to vouch for something but to hypothesize it. 'Suppose it is true', we muse, 'What follows?' 'What he says is either true or false: if it is true, he is a fool; if it is false, he is a rogue; either way we cannot trust him.' 'Suppose, for the sake of argument, that there is no such point: then such and such follows, which is a self-contradiction; so there must be such a point.' In propositional calculus arguments by dilemma and by *reductio ad absurdum*, are equivalent to the principle of double negation, that not-not-p implies p. That principle is not a requirement of consistency, and there are intuitionistic calculi in which it does not hold. But it is a requirement of ordinary two-sided argument between two different people arguing different sides of a case. If one makes a claim and the other rejects it, the former needs to be able to say 'No. That's not so' in order to reinstate his original claim. The words 'true' and 'false' are naturally used in such encounters to characterize or evaluate the claims and counter-claims being put forward. It is natural then to seek a certain 'negation symmetry' by which p shall *be* False iff (short for 'if and only if') not-p *be* True. And though we can, as we shall shortly see, have more than two truth-values, there is not only an obvious economy in having only two, but a deeper requirement of absolute consistency, which insists on there being a crucial distinction between theorems and non-theorems, and therefore in cases where there are more than two truth-values picks out some as 'distinguished' while lumping the rest together as undistinguished. These considerations militate against our admitting truth-gaps, and press the concepts of truth and falsity to be not only mutually exclusive, but jointly exhaustive, and lead us to formulate the Principle of Bivalence.

The Principle of Bivalence is one of several different Either-Or principles which need to be distinguished. There is first the so-called Law of the Excluded Middle that p and/or not-p is a tautology, in symbols $\vdash p \vee -p$. This is a purely syntactical formulation, in terms of the sentential connectives 'Either . . . or', and 'not', or $\vee$ and $-$. There is no mention of truth or any other interpretations of the formalism: $p \vee -p$ could be regarded as a well-formed formula of an uninterpreted calculus, which was a theorem simply because it could be derived from the rules of inference. The Principle of Bivalence, by contrast, is formulated in semantic terms.[34] It lays down that every proposition or propositional function *be* either True or False. This is not a tautology nor a theorem in a system of propositional calculus, but a metalogical stipulation about how a propositional calculus is to be interpreted. We could put it slightly formally,

for all p, either p *be* True or p *be* False.

It is normally taken as obvious that a proposition *be* False if and only if its negation *be* True, but since what is in issue is the possibility that a proposition might *be* not True and yet not *be* False, it is as well to make the assumption explicit. If

---

[34] The distinction between the syntactic Law of the Excluded Middle and the semantic Principle of Bivalence was first drawn by Chrysippus. See Cicero, *De Fato*, 16, 38. In the modern world it is due to Jan Lukasiewicz, 'On Determinism', tr. Storrs McCall, *Polish Logic*, Oxford, 1967, pp. 36–7; reprinted in Jan Lukasiewicz, *Selected Works*, ed. L. Borkowski, North Holland, Amsterdam, 1970, pp. 124–6. For a history of the Principle of Bivalence, see pp. 176–8.

we do, we can then substitute for the word 'False' in the Principle of Bivalence, and obtain, as an alternative formulation

for all p, either p *be* True or not-p *be* True.

This is sometimes called the Semantic Law of the Excluded Middle, and by its resemblance both to the Principle of Bivalence and to the syntactic Law of the Excluded Middle, stated above, shows their resemblance to each other.

It is useful to view the Principle of Bivalence formally as a requirement on valuation functions, in this case a function from propositions into some set of truth-values, such as the two-element Boolean ring, B_2, {TRUE, FALSE}. If it is a *total* function, then every proposition has one of the values, TRUE or FALSE: if it is a partial function, then not every proposition need have a truth-value; although some do and have either the value TRUE or the value FALSE, some may not have either. The Principle of Bivalence is then a requirement that the valuation function be total rather than partial, that is to say, that it be from propositions not only *into*, but *onto* the two-membered set {TRUE, FALSE}. It is a perfectly possible stipulation for a logician to lay down, though it carries consequences for the meaning of the other terms involved – the propositions and the terms 'TRUE' and 'FALSE'. If we start with a partial function into {TRUE, FALSE}, and decide to make it a complete function, we can do so in three ways. We may cut down the field of the function, and disallow those apparent propositions to which neither TRUE nor FALSE can be assigned, and say that they are not really propositions. Alternatively we may create a new value in the range of the function – in this case having a third truth-value – and say that those propositions which neither *have* the truth-value TRUE nor *have* the truth-value FALSE, shall *have* the third truth-value, which may be called NEITHER, or NEUTER, or INTERMEDIATE. Or thirdly, we may achieve our purpose of turning a partial function into a complete function by allocating to all the propositions not yet possessing a truth-value one of the available ones.

It is easy to confuse the original use with the second of these courses: on the one hand a partial function *into* a two-membered set of truth-values; and on the other hand a total function *onto* a three-membered set of truth-values. In each case it would be natural to say that every proposition *be* True, or *be* False, or *be* Neither. But in the former case we are saying that it not *be* either, that is to say that the proposition *do not have* either of the possible truth-values, whereas in the latter we are saying that it *be* Neither, that is to say that the proposition *do*, indeed, have a truth-value, but a different one from either TRUE or FALSE, In the one case we are merely denying the Principle of Bivalence: in the other we are positively asserting a Principle of Trivalence, positing a total valuation function onto a three-element Boolean ring, B_3, whose members are TRUE, FALSE and NEITHER.

We thus have a schema of four possible responses to the demand for bivalence: we may simply reject it; or we may seemingly accept it by the Procrustean policy of not recognising as a proposition any entity that does not fit the principle; or we may accept that every proposition should *have* a truth-value, and define a new one to accommodate difficult cases; or we may insist that every proposition shall *be* True, or else *be* False, and accept that there will be some some shift of meaning to take account of this.

It is difficult to be brusque and to reject the demand for bivalence completely. Though there are difficult cases which seem to stand in the way of our accepting the Principle of Bivalence unreservedly, it has commended itself to many philosophers from Aristotle onwards, and seems to be constitutive of at least some sorts of truth.

The second response is to disallow those apparent propositions to which neither TRUE nor FALSE can be assigned, and say that they are not really propositions – if I talk about the present King of France or Ryle's youngest son, you will check me and say 'But there is no King of France' or 'Ryle did not have any sons' and argue that my utterance, although in propositional form, was not really expressing a proposition because of a failure of reference. In this spirit we might contend that since the future did not exist, future contingent utterances did not really, despite appearances, express propositions. But although it is a possible response for an ordinary language philosopher in view of the linguistic discomfort we have about the future tense and ascriptions of future truth, it is an unattractive course. It seems absurd to assimilate the truth of utterances about the future to the non-truth of statements about the present King of France or Ryle's youngest son. We are impelled to allow that utterances can in principle be said to be true or to be false, even if not accepting that they must always be one or the other. If some predictions about the future can be said to be true, we cannot save the Principle of Bivalence by claiming that no utterance in the future tense expresses a proposition.

The third response was originally put forward by some of the Schoolmen and has been revived in this century. It postulates a third 'intermediate' or 'neutral' truth-value. It is often discussed as though it were a question of fact – metaphysical fact, perhaps, but fact none the less. But it is not so much a fact as an option, and it is important to see it only as one option among others. Our usage is not fixed, and gives no firm guidance on how we should talk. Rather, it is a matter of stipulation, and there are different sets of rules we can lay down, with different advantages and disadvantages. Our chief tasks are to ensure that each set of rules is internally consistent, and that different sets of rules are not confused with one another. If we allow that there shall be three truth-values, then we can represent indefectible truth fairly faithfully, and assign to propositions of the RE form sometimes the truth-value TRUE, sometimes the truth-value FALSE, and sometimes the truth-value NEITHER. It is clearly a possible assignment: after all we assign probabilities on this pattern, sometimes assigning 1, sometimes assigning 0, and sometimes assigning a probability in between. But when we do this for probabilities, we find that the rules for the sentential connectives are not truth-functional: we can add probabilities to form the probability of their disjunction only when they are mutually exclusive; we can multiply them to form the probability of their conjunction only when they are independent; and the rule for conditional probabilities is quite unlike anything that the equivalence between $p \rightarrow q$ and $-p \vee q$ in propositional calculus would lead us to expect. Similarly, attempts to give truth-functional definitions of −, ∨ , and &, in terms of three truth-values, TRUE, FALSE and NEITHER have always run into great difficulties especially with the truth-values to be assigned to $p \vee -p$ and $-(p$ & $-p)$.[35] Only in the extreme case, when there are just the two truth-values, does truth-functional definition even begin to look plausible, and then only in a few instances. To replace the Principle of Bivalence by a Principle of Trivalence or a Principle of Multivalence is something we may be led to in some areas of logic, but it does not meet the requirements that originally led us to postulate the Principle of Bivalence.

The fourth and final option is to insist on the Principle of Bivalence and to allocate one or the other of the two truth-values, TRUE or FALSE, to the propositions which thus far *do not have* one of those truth-values, so that every proposition and propositional content shall *have* a truth-value, and only two truth-values, TRUE and FALSE, are admitted. Locutions which in ordinary discourse we hesitate to dub true or to dub false shall none the less be deemed to be one or the other. Since we need a rule to apply in all cases, we must decide either to deem all the dubious ones True, or to deem them all False. Although either rule is possible, the natural course for the logician to take is the 'fail-safe' one, that anything which be not True is to be regarded as False.

Such a stipulation can perfectly well be made, and there may be good reasons for making it. But it cannot just be made. If made, it will exact a price in the form of alterations to the significance of some of the terms involved. If we say that all those not yet assigned a truth-value are to be deemed to have the truth-value FALSE, we are strengthening the meaning of 'TRUE' and giving it a sense of 'definite Truth', or 'necessary Truth'. A representation of temporal truth that satisfies the Principle of Bivalence will become also a representation of temporal necessity. In thus altering the meaning of the

[35] See further below, ch. 5, §(ii), pp. 91–2.

predicates 'True' and 'False' we pay a further logical price. We have to give up symmetry under negation, which was one of the considerations that led us to require the Principle of Bivalence in the first place. Normally we have the equivalence:

'*p*' *be* False if and only if 'not-*p*' *be* True.

But if we insist on counting as False everything that *be* not True, we can no longer keep this equivalence. Often we would hesitate to say that either '*p*' or 'not-*p*' was true, and so would at first be inclined to say neither that '*p*' *be* True nor that 'not-*p*' *be* True. If this hesitation to ascribe Truth is taken as an ascription of Falsity, then we shall be quickly led to the inconsistency of saying both that 'not-*p*' *be* True and that 'not-not-*p*' *be* True.

In ordinary discourse we value negation symmetry more than bivalence. We may therefore seek to alter not the meaning of the truth-values, TRUE and FALSE, but that of the propositions themselves. Whereas this is hardly possible with bare propositional contents, with more complicated propositions containing temporal standpoints it is feasible, and does in fact happen, importing a modal element into our understanding of tenses.[36] Since modal considerations are here paramount, and in any case loom large in our view of the future, I must now digress to give a metalogical account of modal logic.

[36] See below, ch. 8, §(v), p. 153.

5

Modal Logic

(i) Syntactic Approach

We often modify statements, and there are many modes of discourse besides simple declarative fact-stating sentences. Besides conveying to you the fact of Brutus' killing Caesar, I may qualify it with a 'perhaps', or say that I think it, or claim that I know it, or say that it is definitely true, or indubitably so. I may give advice, and tell him that he should, or pass judgement and say that he ought not to, or give vent to wishes and sigh 'Oh, that Brutus would kill Caesar' or 'Oh, that he had not'. I can outline possibilities, or recognise necessities, or consider counterfactual conditionals. I can engage in fiction and tell stories, or consider obligations, or distinguish the conjectural from the well-established. I can hope, expect, fear, warn, promise, or threaten, about things to come, and can remember, ponder, relate, or explain, the past. All these activities have some propositional content – we can say what the content of our hopes, wishes, judgements or romances is, and pick out entailments and inconsistencies among them – but cannot be represented in terms of propositions alone.

Grammatically, modification of a statement is often expressed in English by the use of an auxiliary verb, and in inflected languages by a change of mood or tense. But whatever the shift of tense or mood, and whether it is expressed by an auxiliary verb or some more complicated locution, such as 'it is possible that . . .' or 'it was going to be the case that . . .', the modified statement stands in some relation to the original one, and is still something that can be agreed with or disagreed with, accepted or rejected, shared or repudiated. It is reasonable to regard the modified statement as a function of the original statement, and it is not too great a departure from ordinary usage to regard the modified statement as still being itself a statement of some sort (perhaps of a somewhat extended sort – so as to include wishes and warnings). It is therefore reasonable to see the various modifiers as unary operators (or functors, or connectives) which operate each on a single proposition to yield a single proposition. The operator 'I dreamt that' converts the proposition that there *be* a sea battle on February 29th, 2000, into the proposition that I dreamt that there was one.

Logic is not much concerned with particular propositions, but abstracts from them to general propositional form. It therefore considers not particular propositions, but propositional variables and other well-formed formulae constructed from them. It is necessary, though often difficult, to distinguish well-formed formulae from particular propositions. Propositional variables are traditionally represented by p, q, r, p', p'', etc., and I shall follow that usage. Where there is a subscript, $p_{\text{February 29th, 2000}}$, or p_{29} or p_u,[1] referring to a definite date, February 29th, 2000, or tomorrow, it represents a proposition, that there *be* a sea battle tomorrow, February 29th, 2000. But in accordance with the (somewhat sloppy) practice of mathematicians, alphabetical subscripts will also be used on occasion to represent variables. Granted a rule of substitution, which is a feature of most logical systems, we can make do with just propositional variables, but often it is convenient, especially in the syntactic approach, where we are concerned with well-formed formulae apart from any interpretation we might give them, to represent them differently. Often capital Greek letters are used, or capital A and B: but I have used B for a Boolean ring, and shall use A for the universal quantifier, and shall soon be suffering from a scarcity of alphabetical resources. I shall therefore, with apologies to the professionals, use the capital

---

[1] See above, ch. 4, §(ii), pp. 63–5; ch. 2, §(iii), p. 24; Preface, p. x.

letters *C*, *J*, and *U*, for well-formed formulae but make use of *p* and *q* where possible. The general form of a logic of moods and tenses will be that of the propositional calculus enriched by a further unary operator, generally represented by a small square, □, or a capital bold **L**, with the same formation rules as for negation, and appropriate rules of inference and axioms.[2] Our task then, on the syntactical approach, is to consider what theorems follow, granted particular axioms and rules of inference, and in the light of these to see what axioms and rules really are appropriate.

The unary operator may stand for any one of a wide variety of modal or tense auxiliary verbs or adverbs such as 'perhaps', 'possibly', 'certainly', or propositional phrases such as 'It is known that . . .', 'I know that . . .' 'It is said that . . .', 'Homer relates that . . .', 'Let's pretend that . . .', 'It was the case that . . .', 'It will be the case that . . .' and 'It is the case that . . .'. Different modal operators will require different axioms. It is not to be assumed in advance that every modal operator in ordinary speech can be expressed adequately in modal logic: sometimes modal logic will reveal ambiguities in ordinary usage we are unable to resolve. But it is a useful exercise to see how far we can go in considering the formal possibilities enriching propositional calculus by a simple basic unary operator, and the constraints on the rules of inference and axioms it is reasonable to recognise. Although this is undertaken as a formal exercise it is not only that. Formal constraints exert some pressure on ordinary linguistic practice. Admittedly, we do wrong to straitjacket ordinary usage within the Procrustean rules of formal modal logic, and we need to be sensitive to the nuances of actual language, but we shall understand actual usage better if we are aware of what possibilities there are available which do not on the one hand

[2] In this book the sentential connectives will be represented as follows: conjunction by &, disjunction by ∨, implication by →, biconditional by ↔, negation by —, and entailment by ⊢; conjunction is often represented in older books by . , and in modern ones by an inverted wedge, ∧; and material implication by ⊃.

run into inconsistency nor on the other peter out into vacuity. With this warning, therefore, we give a syntactic exposition of an enriched propositional calculus and the systems which arise naturally from the constraints inherent in the enterprise.

Having added a unary operator, □, to propositional calculus with its formation rules, we need to consider possible rules of inference and axioms governing the use of □. There is quite a wide range of possibilities, and we shall find that there are many modal logics. But there are certain constraints. If we have too many rules or axioms, the modal operator will be degenerate, definable in terms of the ordinary sentential connectives, and modal logic will be nothing more than ordinary propositional calculus: if we have too few, the modal operator will lose all contact with the connectives of ordinary logic, and modal discourse will be no longer a logic at all. We must steer a careful course between the Scylla of having too many rules or axioms, with the result that the modal operator is vacuous or gives rise to inconsistency, and the Charybdis of having too few with the result that our modal discourse is a chaotic one in which none of the ordinary rules of ordinary discourse apply.

Many operators are chaotic. 'I dreamt that . . .', 'It is said that . . .', prepare us for a complete absence of coherence or rationality. But if we are to give sustained attention to a mode of discourse, some conditions of intelligibility must be satisfied. If words are to have their ordinary meanings, analytic propositions must hold as well within the modalised discourse as outside it. As far as propositional calculus is concerned, tautologies must remain tautologies when modalised. Since every tautology is a theorem of propositional calculus, and *vice versa*, we stipulate

$$\text{If } C \text{ is a theorem, so is } \Box C,$$
i.e.,   If $\vdash C$ then $\vdash \Box C$.

This rule of inference is known as the **Rule of Necessitation**, abbreviated as **R□** or **RL**, and is characteristic of all modal logics. But even with this rule of inference we need to be wary. It would not be valid if □ were interpreted as 'I know that' and

⊢ C were a theorem of, say, number theory or advanced logic. Nevertheless, we lay it down as an acceptable idealisation, and note that it is very similar to the Rule of Generalisation in predicate calculus

$$\text{If } \vdash C \text{ then } \vdash (Ax)C.^3$$

The Rule of Necessitation ensures that logical theorems remain so when modalised, but does not by itself suffice to legitimise standard inferences in modalised discourse: it enables us to introduce an entailment ⊢ into a mode of discourse, but not to use it to make inferences within it. If we are to carry ordinary inferences over into modalised discourse we need the further rule

$$\text{If } p \vdash q \text{ then } \Box p \vdash \Box q.^4$$

Modern formal logic, in order to distance itself from substantial issues about the validity of inferences in particular disciplines, which depend on the way the world is, or on various principles of legal, political, historical, or philosophical argument, tends to cast all inferences, other than those which are simple substitutions or which can be shown to be purely tautological, into the *Modus Ponens* form

$$p \to q, p \vdash q.$$

In this way the question is moved from a question of validity, whether $p \vdash q$ is a *valid inference*, to a non-logical question of truth, whether $p \to q$ is a *true implication*.[5] And so the issue whether modalised inferences are real inferences, that is

[3] In this book the universal quantifier will be represented by (Ax); in older works on formal logic it is usually represented by (x), and in many modern ones by (∀x), with an inverted A. In almost all books the Existential quantifier is represented by (∃x), with a reversed E, but in this book it will be written (Vx). The advantage of this, apart from ease of typography, is that it suggests the parallel between the universal quantifier and the ∩ for intersection, the cap ∩ and the modern ∧ for conjunction, and the parallel between the existential quantifier and the ∪ for union, the cup ∪, and the ∨ for disjunction.

[4] Aristotle, *Prior Analytics*, I, 15, 34ª 22–4.

[5] But see below, ch. 6, §(i), pp. 91–2 of 14.6.88.

whether $\Box p$, $\Box(p{\rightarrow}q) \vdash \Box q$, becomes the question whether a modalised implication $\vdash \Box(p{\rightarrow}q)$ yields a straightforward implication between the modalised parts of it, $\Box p$ and $\Box q$. We therefore lay down as an essential axiom for modal logic

$$\text{K} \qquad\qquad \Box(p \rightarrow q) \rightarrow (\Box p \rightarrow \Box q)$$

We should note once again the parallel between Axiom K and the axiom for predicate calculus

$$(\text{Ax})(p \rightarrow q) \rightarrow ((\text{Ax})p \rightarrow (\text{Ax})q)$$

The axiom K entitles us to infer $\Box q$ from $\Box p$ and $\Box(p{\rightarrow}q)$ in two steps of *Modus Ponens*. It was on account of this that we could have inferred Ls from $L(Gks \rightarrow s)$ had we been given $LGks$, but not if we were given only Gks.[6]

The Rule of Necessitation together with the Axiom k ensure that modalised discourse is '**inferentially transparent**'. Essentially what we require is that we should be able to make the same inferences within modalised discourse as in unmodalised discourse. If there is a good argument about kicking the ball – *e.g.* that in order to kick it, one must approach it, or that if one kicks it, the result will be that it moves – the same argument should hold within the context of obligatory kicking, alleged kicking, future kicking or past kicking. Else modal discourse becomes inferentially opaque. It is worth emphasizing the element of stipulation here. It is not denied that there are unary operators which are inferentially opaque, such as 'I dreamt that . . .' or 'It is stated in *Pravda* that . . .'. No rationality, no consistency even, is expected of dreams. But such a criticism, if made out, is a damaging one. If dreams, or the pronouncements of *Pravda*, are an inferentially opaque universe of discourse within which normal arguments do not apply, then the operator 'I dreamt that . . .', or 'It is stated in *Pravda* that . . .' carries the connotation of not being worthy of serious regard. Hence it is reasonable to stipulate that the modal operators we should be seriously concerned with are ones which allow inferences to be

---

[6] See above, ch. 3, §(iv), p. 42.

drawn. We therefore define the system which satisfies the Rule of Necessitation and Axiom к as being the minimum system of modal logic. We call it **K**.[7] Other systems are those that satisfy further axioms expressing further entailment patterns for the modality in question. But so long as they include к they avoid the Charybdis of logical chaos, and can be regarded not just as modalised discourse, but as modal logic.

The Rule of Necessitation and Axiom к, which together guarantee inferential transparency, govern the interrelation of the modal operator $\Box$ with implication and entailment, $\rightarrow$ and $\vdash$: in order to place it as fully as possible in the context of propositional calculus, we need also to consider its interrelationship with negation, $-$. As a first try, we might think that it would 'commute' with negation, *i.e.*

$$\vdash \Box - p \leftrightarrow - \Box p,$$

but in that case the modal operator would become vacuous, so far as propositional calculus was concerned. To see this, it is useful to be able to cite one theorem of **K**

$$\vdash \Box(p \ \& \ q) \leftrightarrow \Box p \ \& \ \Box q.$$

With the aid of this theorem we can see that

$$\Box - p \leftrightarrow - \Box p$$

must be a non-theorem. For if it were a theorem the modal system would become in a sense degenerate. Suppose

$$\vdash \Box - p \leftrightarrow - \Box p.$$

Then consider any well-formed formula $\Box U$ where U is a molecular well-formed formula of the propositional calculus. Then U can be expressed in terms of negation and conjunction,

---

[7] In order to distinguish between axioms and the logistic systems they are characteristic of, I use small light capitals for the former and large bold capitals for the latter. Thus к is the axiom, and **K** is a system containing only к (apart from the axioms and rules of propositional calculus and the Rule of Necessitation). For other formulations of the system **K**, see Brian F. Chellas, *Modal Logic*, Cambridge, 1980, §4.1, pp. 114–16.

and must then be either of the form $-C$ or of the form $C \& J$ where C and J are well-formed formulae of the propositional calculus. If U is of the form $-C$, then $\Box U \leftrightarrow -\Box C$; if U is of the form $C \& J$, then $\Box U \leftrightarrow \Box C \& \Box J$: hence for any well-formed formula, U, of the propositional calculus, there is an equivalent one in which every modal operator governs only a single propositional variable; and hence for any well-formed formula whatsoever of the form $\Box U$, there is an equivalent one consisting of negations and conjunctions of either atomic propositional variables or atomic propositional variables governed by one or more modal operators. We can, as it were, nest all the modal operators in the innermost parts of a well-formed formula of propositional calculus. $\Box p$ does not, typically, have the same truth-values as p, nor do its truth-values depend on p. But it does have some truth-values, and can therefore be represented by some other propositional variable, say p'. We should be simply writing $\Box p$ for p', $\Box\Box p$ for p'', *etc.*, if we had

$$\vdash \Box - p \leftrightarrow -\Box p,$$

as a theorem. It therefore cannot be a theorem, if the system is not to degenerate into a much simpler one. Such a degeneracy may not always be bad. Indeed, in chapter 9 we shall have to consider modalities that do commute with negation. But there is good reason why in general they should not, and why we should regard those that do as being in some way untypical.

The non-theorem

$$\Box - p \leftrightarrow -\Box p,$$

consists of two conjuncts

$$\Box - p \to -\Box p \text{ and } -\Box p \to \Box - p.$$

Although we cannot have both, we can, and should hope to, have one; else our modal operator $\Box$ or L will have so little to do with the ordinary sentential connectives, i.e. $-$, $\to$, $\leftrightarrow$, $\&$ and v, that there will scarcely be a logic worth talking about. Although we could choose either, and the decision is, as we shall shortly see, in some sense arbitrary, we shall choose the former conjunct $\Box-p \to -\Box p$. The reason is that we naturally want to secure a certain 'modal consistency' for our operator. One standard formulation of the ordinary requirement of consistency is that no well-formed formula of the form $p\&-p$ can be a theorem. We naturally go further — though it is further — and lay down that not only is $p\&-p$ not a theorem, but that the negation of $p\&-p$ is a theorem, that is,

$$\vdash -(p\&-p).$$

This is a theorem of ordinary propositional calculus. In considering the relation between $\Box$ and $-$, we may reasonably look for a comparable stipulation, *viz.* $\vdash -(\Box p \ \& \ \Box-p)$. By standard propositional calculus this is equivalent to

$$\Box-p \to -\Box p,$$

the former of the two conjuncts. This in turn is equivalent to

$$\Box p \to -\Box-p,$$

which is a characteristic thesis of modal logic, known as the axiom D. Although the minimum modal system, **K**, does not have D as an axiom but only κ together with the rule of necessitation R$\Box$ or RL, almost all interesting systems do.

There are good reasons for accepting axiom D, but they are not coercive. We can, if we wish, adopt the other conjunct as an axiom, provided we make corresponding adjustments to the other axioms. If we do, we define a different, weaker modality, which we write $\Diamond$ or **M**, and have

$$\vdash -\Diamond-p \to \Diamond p.$$

The similarity between these two forms is suggestive. The operator $-\Box-$ behaves like $\Diamond$, the operator $-\Diamond-$ behaves like $\Box$. If we adopt either as basic, we can with the aid of negation, define an operator with the properties of the other. Some writers have taken **M** as basic, and defined

$$\mathbf{L}p =_{\text{def.}} -\mathbf{M}-p, \text{ that is, } \Box p =_{\text{def.}} -\Diamond-p,$$

and developed an eqivalent modal logic on that alternative basis.[8]

Nevertheless, there is a difference between the two operators. The argument from inconsistency is weighty. D is equivalent, we have seen, to $-(\Box p \,\&\, \Box -p)$, which is analogous to the Law of Non-Contradiction $-(p \,\&\, -p)$. If $\Box p \,\&\, \Box -p$ were allowed ever to be asserted in modal discourse, then by virtue of K we could also assert $\Box(p \,\&\, -p)$. That is, the rationale of $\Box p \to -\Box -p$ is that it blocks the route to asserting an inconsistency within modal discourse. The corresponding postulate for $\Diamond$, namely $-\Diamond p \dashrightarrow \Diamond -p$, transforms into $-(-\Diamond p \,\&\, -\Diamond -p)$ and into $\Diamond p \vee \Diamond -p$. These do not block any inconsistencies, and so guarantee no inferences. Moreover, weak modalities are not strongly inferential. Although

$$\text{from } \vdash p \text{ we can infer } \vdash \Diamond p,$$

we do not have

$$\Diamond(p \to q) \to (\Diamond p \to \Diamond q)$$

as a theorem. The fact that we have $\Box p \to \Diamond p$ as a theorem, but not $\Diamond p \to \Box p$ suggests that $\Box$ is inferentially stronger than $\Diamond$. The fact that we have $\Diamond p \vee \Diamond -p$ as a theorem, but not $\Box p \,\&\, \Box -p$ suggests that $\Box$ is inferentially more stringent than $\Diamond$. The fact that we have $\Box(p \to q) \to (\Box p \to \Box q)$ as a theorem, but not $\Diamond(p \to q) \to (\Diamond p \to \Diamond q)$ suggests that $\Box$ is inferentially more transparent than $\Diamond$. Hence, although it is formally possible to take weak modalities as basic, it is conceptually preferable to take strong modalities as basic, as they are inferentially stronger, more stringent, and more transparent.

If we add the axiom D to the system **K**, we obtain the system **D**. In the system **D** we have four out of the six possible interconnexions between $\Box$ on the one hand and $\&$, $\vee$ and $-$ on the other, *viz.*

---

[8] For example, G. H. von Wright, *An Essay in Modal Logic*, Amsterdam, 1951.

$$\Box(p \ \& \ q) \to \Box p \ \& \ \Box q$$
$$\Box p \ \& \ \Box q \to \Box(p \ \& \ q)$$
$$\Box p \lor \Box q \to \Box(p \lor q)$$
$$\Box -p \to -\Box p$$

but not

Necessity Divided under Disjunction $\Box(p \lor q) \to \Box p \lor \Box q$, which would legitimise the inference from *Every natural number must be either even or odd* to *Either every natural number must be even or every natural number must be odd*,[9]

nor

Commutativity with Negation D* $-\Box p \to \Box -p$, which would make negation and the modal operator completely commutative;

and we cannot add either of these on pain of modal degeneracy.[10] We can therefore argue for the system **D** as giving us as much, in the way of interconnexion between the modal operator $\Box$ and the connectives &, $\lor$ and $-$ of propositional calculus, as we can hope to have.

Most modal logics have D as an axiom, but in tense logic it precludes various possibilities we do not wish to foreclose by *fiat*, and we do not make it part of the definition of a modal logic that it should include D. It is for this reason that the minimal modal logic **K** was defined solely with regard to $\vdash$ and $\to$, by means of the Rule of Necessitation and the axiom K. D is one step more fully integrated with propositional calculus, in that the relation between $\Box$ and $-$ is considered as well as that between $\Box$ and $\vdash$ and $\to$. We can think of the propositional calculus as a Boolean Algebra on a set X with one unary and

---

[9] See further below, ch. 7 §(i), p.122.
[10] See below, ch. 9, §(ii) and §(iv), pp. 165, 175, where it is a test for the modal vacuity of some tense operators that they commute with negation or divide under disjunction.

one primitive binary operator, together with one other non-primitive binary operator, the whole algebra usually written ⟨X, , ,′⟩. Having explored this simple system, we want to go a stage further. The smallest step to take is to add a unary operator, such as □, and consider ⟨X, , ,′,□⟩. We then consider possible interrelations between this unary operator and the other operators, and argue that our rules for □ make it the **most highly structured non-trivial operator relative to the Boolean operators**.

There is a parallel with topology. We can see topology as an enrichment of the Boolean algebra of sets. In standard expositions of topology, we pick out a family of sets – the open sets – which are distinguished by certain infinitistic properties: infinite unions, though only finite intersections, of open sets are always open. We can, however, axiomatize topology in terms of a unary operator, the interior operator, which is a function from sets into sets, just as □ is a function from propositions into propositions; and then we find a close parallel between the axioms of topology and those of a particular modal logic, S4.

Although the system **D** goes as far as possible in relating the modal operator with the sentential connectives of propositional calculus, it leaves other questions unanswered. It tells us nothing of the relation between modalised and unmodalised discourse, nor of any relations between iterated modal operators. Axioms giving rules for such relations can be laid down, giving rise to different modal logics, according to what rules are adopted. The axiom т, giving rise to the system **T**, is an example of the former, the axioms **4** and **5**, which are typical of the systems **S4** and **S5**, are examples of the latter; the axiom в, which is typical of the system **B**, is an example of both.

(ii) Semantic Approach

Modal operators can be interpreted in many different ways. The only one excluded is a truth-functional one with only two truth-values; for then the modal operator could be defined truth-functionally, and modal logic would degenerate into a version of propositional calculus. But an interpretation in terms of a many-valued logic is permissible, as foreshadowed in the previous chapter.[11] If there are three truth-values, TRUE, FALSE, and NEITHER, then the following truth-functional interpretations of $\Box$ and $\Diamond$ suggest themselves.

p	$\Box p$	$\Diamond p$
T	T	T
N	F	T
F	F	F

Here, essentially, $\Box$ cuts out the penumbra of uncertainty, concentrating on only what *be* definitely True, while $\Diamond$ indicates the penumbra, and excludes only what *be* definitely False.

The difficulty comes when we try to give analogous definitions of the sentential connectives. Although there is no difficulty with negation, which can be given a truth table thus:

p	$-p$
T	F
N	N
F	T

there are difficulties with implication, conjunction and disjunction. It seems natural to stipulate that N&N *be* N, and likewise NvN *be* N. But then $p\&-p$ *be* not in all cases False, and $pv-p$ *be* not in all cases True. We have noted similar difficulties arising in the calculus of probabilities, which can be seen as a many-valued logic in which there are a non-denumerable infinity of truth-values, all the real numbers from 0 to 1. In the calculus of probabilities we cannot in general

[11] ch. 4, §(iv), pp. 76–7.

give a truth-functional account of conjunction, except under the condition that the two conjuncts be independent, or of disjunction, except under the condition that the two disjuncts be exclusive.

These difficulties need not be taken as objections to the semantic approach, but rather as indicating limitations on truth-functional definitions in many-valued logic, which parallel the difficulties about material implication, familiar even in two-valued logic. Material implication, $\rightarrow$, can be defined truth-functionally, but does not correspond to our ordinary notion of implication, which requires not just that the antecedent *be* False or the consequent *be* True, but that there be some further connexion between them. Truth-functional definitions are in general too crude to express the full subtlety of logical terms such as *if ... then ...*, *and*, and *and/or*, and when we come to consider the interpretation of $\square$ which assigns TRUE to $\square p$ if *p have* the value TRUE, and FALSE otherwise, we find that although it expresses well what we mean by 'definitely true' or 'definitely false', it cannot accommodate the fact that though it is not definitely true that there will be a sea battle tomorrow nor definitely true that there will not be one, it is definitely true that there either will be or will not be one.

(iii) Kripke Semantics

Another semantics for modal logic multiplies not truth-values but possible worlds, exploiting the parallel, already noted, between the $\square$ operator and the universal quantifier (Ax). Leibniz suggested that what *be* necessarily true *be* that which *be* true in all possible worlds, and Kripke refined that interpretation by considering not all possible worlds but all accessible worlds, where accessibility is defined by some relation Q. Different sorts of relations provide models of different sorts of modal logics, expressing different senses of the words 'necessary' and 'possible'.[12]

[12] See above, ch. 3, §(iii), p. 38.

If Q is reflexive, we have a model for the modal logic **T**, in which we have as an axiom

T $\qquad\qquad \vdash \Box p \to p.$

If Q is symmetric, we have a model for the modal logic **B**, in which we have as an axiom

B $\qquad\qquad \vdash \Diamond \Box p \to p.$

If Q is transitive, we have a model for modal logics like **S4**, in which we have as an axiom

4 $\qquad\qquad \vdash \Box p \to \Box \Box p,$

or, equivalently, as we can show, by substituting $-p$ for p, and contraposing,

4$''$ $\qquad\qquad \vdash \Diamond \Diamond p \to \Diamond p.$

If Q is an equivalence relation, we have a model for the modal logic **S5**, in which we have as an axiom

5 $\qquad\qquad \vdash \Diamond \Box p \to \Box p.$

These correspondences are illuminating, and the connexions between different sorts of relation give us deep insight into the ways modal logics are connected to one another.[13]

(iv) Prior's Tense Logic

Relations have converses. The relational structure of models of modal logic encourages us to think of the inverse modal operators. We could quite naturally write down $\Box^{-1}$ and $\Diamond^{-1}$ and consider them also as unary operators adjoined to propositional calculus. But we shall not gain anything thereby in **B** and **S5**, and any other system containing **B** as an axiom or theorem, since the accessibility relation Q is then symmetric, so

---

[13] For fuller and illuminating discussion of Kripke semantics, see Chellas, *Modal Logic*, esp. chs. 1 and 3; and G. E. Hughes and M. J. Cresswell, *Introduction to Modal Logic*, London, 1968, ch. 4.

that $\Box^{-1}$ will be just $\Box$ itself. These systems of modal logic cannot capture the use we make of tenses, where the future operator is essentially different from its inverse, the past operator. The requirement imposed by axiom 4, $\Box p \rightarrow \Box\Box p$, that the semantic relation be transitive, is exactly what we want for a tense logic. S4 itself is not entirely apt, in as much as it contains the thesis T, and so includes the present in the future and the past. In the classical development of tense logic by A. N. Prior, unmodalised propositions represented the present tense, and if we had T as an axiom[14], we should be committed to holding that anything true of the future or past was true of the present too. In the RE calculus that I shall develop in the next chapter, all temporal standpoints are explicitly expressed, and there is no danger of assimilating the present to the future or the past inadvertently; but the formation rules will preclude our having T as an axiom, so that though the system is in some respects T-like, they do not actually have T as an axiom or theorem. When however we come to devise semantics it is much less cumbersome to have Q reflexive, and at that stage we shall drop the third requirement.[15] We therefore define tense logics in this chapter as those modal logics for which the accessibility relation is

1)   asymmetric
2)   transitive
3)   irreflexive

and later as those modal logics for which the accessibility relation is

1)   antisymmetric
2)   transitive
3)   reflexive.

In either case the modal consequences are the same, in that tense logics:

---

[14] See below, ch. 6, §(i), pp. 108–9.
[15] See below, ch. 8, §(ii), p. 144.

1) do not contain B
2) contain 4'

and in this chapter

3) do not contain T.

Condition(1) stipulates that the accessibility relation Q is asymmetric (or antisymmetric), and hence that there is a difference between the operator □ and its inverse $□^{-1}$; condition(2) stipulates that the accessibility relation Q is transitive and hence that both the operator ◇ and its inverse $◇^{-1}$ are idempotent, in the sense that the future's future will be itself future, and the past's past itself past; and condition(3) secures that both will be different from the present. We thus have the means to develop modal logic into a logic of tenses that at least begins to capture the basic uses of tenses.

The classical development of tense logic is due to Prior, and we shall follow him in using not the □, ◇, $□^{-1}$ and $◇^{-1}$ of modal logic, but **G** and **F** for the future operators and **H** and **P** for the past operators. **G***p be* True iff *p be* True in every Q-accessible world, and **F***p be* True iff *p be* True in some Q-accessible world: **H***p be* True iff *p be* True in every Q^{-1}-accessible world, and **P***p be* True iff *p be* True in some Q^{-1}-accessible world. To a very great extent the force of the tense operators **G**, **F**, **H**, and **P** is best understood by reference to the properties of the accessibility relations Q and Q^{-1}. In particular, they help to make clear the most important feature of tense logic, as distinct from modal logic, which is the interaction between operators and their inverses. It is clear that if *p* is true in the actual world, there is some Q^{-1} world for which there is some Q-accessible world, namely the actual world in which *p* is true. That is

$$p \rightarrow \mathbf{PF}p.$$

In claiming this we are relying on the basic property of relations that it is always the case that $xQ^{-1}Qx$, or, formally,

$$(\mathrm{A}x)(xQ^{-1}Qx).$$

It likewise is always the case that $xQQ^{-1}x$, or formally,

$$(Ax)(xQQ^{-1}x), \text{ which yields,}$$
$$p \to FPp.$$

Kripke semantics enables us to say more. If p is true in the actual world, Fp is true in every Q^{-1}-accessible world, since for any relation Q in a non-empty universe,

$$(Ax)(Ax)(Vz) \ (xQ^{-1}y \to yQz).$$

So we lay down the stronger axioms

$$p \to HFp \text{ and } p \to GPp.$$

In most tense logics it is sufficient to state the latter pair of theses, since $Gp \to Fp$ and $Hp \to Pp$ in all systems where D and D^{-1} hold. But if there were a last moment of time, Gp would count, somewhat Pickwickianly, as holding then while Fp would not, and similarly Hp would be accounted True at the first moment of time, if time had a beginning, while Pp would be False. None the less, $p \to HFp$ is true, no matter what the truth-value of p or Fp, because if there are no Q^{-1}-accessible worlds, then, in a Pickwickian sense, HFp is true in all of them, and similarly $p \to GPp$. Hence our two axioms are not only usually the stronger, but more widely true, and thus meet to be adopted as axioms.

If we substitute $-p$ for p in the two axioms and contrapose, we obtain two theorems

$$PGp \to p \text{ and } FHp \to p.$$

We should note, however, that if we interchange the operators, we do not have theorems. The theses

$$p \to FHp, \ p \to PGp, \ GPp \to p \text{ and } HFp \to p,$$

are none of them theorems in tense logic, except in systems with some special characteristics.

There is an obvious parallel between the axioms $p \to HFp$ and $p \to GPp$, and between the corresponding theorems, and it might seem that tense logic must have this 'mirror-image property' in all cases. This is not so. It is possible for the axiom D to hold, but not D^{-1}. It is conceivable that time should have a beginning but no end, in which case it would be false that $p \rightarrow \mathbf{PF}p$ but true that $p \rightarrow \mathbf{FP}p$. Whether a tense logic possesses the mirror image property with respect to a particular thesis depends on whether the converse of the accessibility relation has to possess the property possessed by the accessibility relation that secures the truth of the thesis in question. If a relation is transitive, symmetric, or reflexive, then its converse is too, and likewise if it is transitive and antisymmetric; so also if it is dense, discrete, or continuous; but not necessarily if it is serial, nor if it is many-one. If we accepted D, as we were inclined to do on purely syntactic grounds,[16] we now see that we should be committed to the accessibility relation, Q or its converse Q^{-1}, being serial, and hence to there being either not a last or else not a first moment of time. But some thinkers have contested this. Some theologians have held that time was created by God, and that the first moment was that at which He created it. Some philosophers have thought that time is only a concomitant of change, and that before the universe began, when there was nothing to change, there was no time: time on their view, only began with the Big Bang. So too some theologians have spoken of the end of time, τὸ ἔσχατον (*to eschaton*), and similarly philosophers have argued that at the Big Crunch, when everything is squeezed out of existence, time too will come to an end. These contentions are questionable, but need to be settled by argument, not *fiat*. We therefore do not stipulate that either D or D^{-1} be a thesis of tense logic, in spite of their syntactic attractions.

[16] See above, §(i), pp. 88–90.

(v) Critique of Prior's Tense Logic[17]

Thus far we have been considering tense logic simply as an amplification of some system of modal logic. We have called it tense logic on account of the fact that temporal relations are asymmetric, and of various parallels in consequence between the extended modal operators and certain temporal expressions. Thus, when I speak of 'the future' or 'the past', a quantificational account is appropriate. I am likely to be referring to all the future times, or all the past times, relative to the time of speaking or relative to some particular person or event. So too, when I say 'Socrates is mortal' or 'Socrates is immortal', I am saying in the former case that Socrates will die at some time, and in the latter that Socrates is always going to be alive. With ordinary tensed expressions, however (except for some uses of the perfect tense[18]), the parallel is far from complete, and we must not take it for granted that Prior's system of extended modal logic accurately represents our normal tense usage.

When we say 'Peter came' or 'Peter will come', we do not naturally mean merely that there was a time when Peter came or that there will be a time when Peter will come. Rather, we have a particular time in mind, and we mean that Peter came then or that Peter will come then. Of course, it follows from Peter's coming at a particular time that he comes at some time. But if I say that Peter will come and you deny it, you are not committed to the claim that Peter will never come, but only to his not coming at the particular time I had in mind. If the correct analysis were in terms of an existential quantification over time, then its negation would have to be in terms of universal quantification over time. But in fact it is not. 'Peter will come' and 'Peter will not come'

---

[17] Jeremy Butterfield, 'Prior's Conception of Time', *Proceedings of the Aristotelian Society*, 84, 1983–84, pp. 193–209, gives a fuller discussion of different aspects of Prior's thought and their bearing on contemporary disputes in philosophical logic.

[18] See below, ch. 6, §(i), p. 105, n. 5.

refer to the same time, which is understood from context, and which could have been referred to explicitly but was not. Neither proposition should be explicated in terms of quantifiers, but rather of a temporal reference which has not been completely filled out.

This means that tense operators in ordinary uses, like in 'Peter will come', are to be elucidated in terms of incomplete specification rather than existential quantification. There is a blank to be filled in, specifying when Peter will come, not an assertion that there exists a time when he will. If the blank is filled in, as in 'Peter will come on Tuesday, May 30th, 2000, at 5.15 p.m.', we have a complete specification of what we are saying. In order to appreciate the logic of tense auxiliaries, we should first consider that of completely specified tensed statements, and then look out for contextual implications of elliptical locutions.

Prior and his followers always speak of branching time. In part, it is a concession to positivist scruples. Time abstracted from events seems too tenuous to have any real existence. If we identify time with events, we secure the existence of time against all such doubts. But we do not need so desperate a remedy. Thanks to Szpilrajn's theorem, we can always extend a partial ordering to a linear one.[19] That is, if we have a partial ordering induced by the accessibility relation Q between possible worlds, or states of affairs, we can extend it to a linear ordering Q^* such that if, to anticipate the terminology of chapter 8, $w_{jr}Qw_{it}$ then $w_{jr}Q^*w_{it}$. We can, as it were, project from a partial ordering on states of affairs with respect to accessibility to a linear one with respect to dates. Admittedly, Szpilrajn needs the Axiom of Choice to prove his result in the general, infinite case. But it

[19] E. Szpilrajn, 'Sur l'extension de l'ordre partiel', *Fundamenta Mathematicae*, 16, 1930, pp. 386–9. Szpilrajn remarks (p. 389) that there is more than one linear ordering that is an extension of a given partial ordering. In the Special Theory of Relativity different frames of reference give rise to different linear orderings of dates which are all based on one invariant partial ordering of causal influenceability between events.

Modal Logic

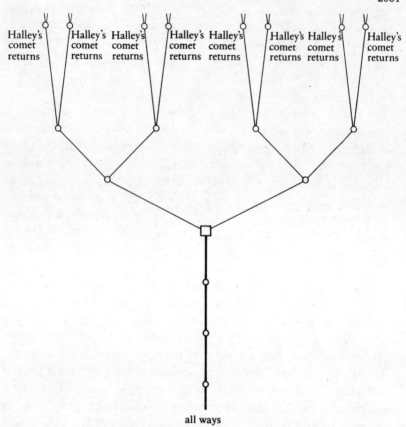

Figure 5:1. Halley's Comet; necessary (all ways). In every possible world at 2061 there *be* Halley's comet.

needs a very sophisticated scepticism to allow infinite numbers of possible worlds and then cavil at the Axiom of Choice.

Quite apart from from being unnecessary, the identification of time with the course of events is open to objections, which I, and others more effectively, have raised.[20] For understanding

[20] J. R. Lucas, *A Treatise on Time and Space*, London, 1973, pp. 9–13. S. Shoemaker, 'Time without Change', *Journal of Philosophy*, 1969, pp. 363–81; quoted W. H. Newton-Smith, *The Structure of Time*, London, 1980, ch. 4, §4, pp. 19–24.

the future correctly it is important to distinguish the different things which may happen in the future from the time when they may happen. There may be a sea battle tomorrow, or there may not: those are different possibilities; but there is only one tomorrow. I may marry Jean in the year 2012 or I may marry Jane: those are different possibilities; but there are not different year 2012s, only different things that may happen in that one year.

The distinction is particularly important when we come to consider determinism, and we need to distinguish 'all ways' from 'always'. Sometimes, whatever I do, and whatever anybody else does, a particular event is bound to take place. Halley's comet will return to be near the earth in the year 2061. There is no way it cannot happen. Whatever way things turn out, it will be so. We can picture the possible courses of events as branching towards the future, and each representing a way forward into the future. And then all ways lead to the same result as far as Halley's comet is concerned. Other things may be different. The US$ may go down with respect to the £ sterling, or *vice versa*. But whatever else happens, Halley's comet will come. It will come all ways. But it will not be always with us. By the winter of 2062–3 it will have gone again, and will not be around for another seventy-six years. Whereas if I go and fall under a bus, I shall go on being dead always. It is not the case that I must fall under the bus. If I take care not to go in cars or wander across main roads, I can make sure that I shall not die in a traffic accident. My dying in a traffic accident is not inevitable, but would be irremediable. My getting hungry in the next day or two is inevitable, but not irremediable. Death, however, is not a good example, because although it is not necessary that I die at a particular time in a particular way, it is necessary that I die sooner or later in some way or another. A better, though more arcane, example is the indelibility of the priesthood in Catholic theology; once a

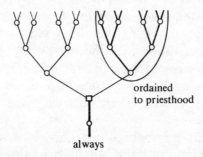

Figure 5:2. Indelible Orders (always). Orders marked solid.

priest, always a priest. I do not have to become a priest, but if I do, I remain one for ever.[21]

Altogether, tense usage differs from Prior's account in three important respects. It does not normally quantify over dates but, rather, refers implicitly to definite dates. It does not assimilate time to possible courses of events. Thirdly and most importantly, it accommodates a mandatory shift of temporal standpoint from one occasion of utterance to another. But this feature of ordinary tensed discourse is compensated for, and often masked, by a fourth, though often unrecognised, feature in that the speaker is free to choose what temporal point of view he wishes to take. Prior accommodates this feature, but does not distinguish it from the third. He allows, that is, our being able to view a proposition from different temporal standpoints, but does not give an adequate account of the way in which what is present now will be past tomorrow.

These considerations will be elucidated in the next chapter. Priorean tense logic, in spite of these criticisms, remains important and profoundly illuminating. It reveals some of the

[21] It is a pity that the favourite example of philosophers, that All men are mortal, is a seemingly simple, but logically very complex, proposition. It is true of me, and of you, and of everybody else, that **all** ways I *die* at **some** time, and thereafter *be* dead **always**. But we need to distinguish the inevitability of our dying sometime both from our having to die at some particular time and from the irrevocability of death when it comes.

basic constraints which govern our use of modifiers which indicate mood or tense, the possible interplay between modal and tense logics, and the way in which the different properties of the underlying accessibility relation determine the particular modal or tense logic. Some properties of models are not easily expressed in syntactic terms – e.g. non-circularity, non-linearity, irreflexiveness of the accessibility relation – and often we find ourselves guided by strong intuitions about models. Although formally we have been considering simply a generalisation of modal logic, our real concern is to give a coherent account of time, and time has many properties besides those required for tense logic. We shall, furthermore, find ourselves increasingly led to invoke new concepts about models not yet explicated, such as the actuality of the present and the connectedness of time.

6

The Logic of Temporal Standpoints

(i) Syntax

We conjugate in tenses as well as in persons, and the change of temporal standpoint, like the change of personal standpoint, greatly alters our perspective. The conjugation of concepts has wide application in philosophy. I decide what I shall do, but advise you what you should do, and later apologize (or boast) on account of what I have done, and blame (or praise) you for what you have done. Moreover, as we saw in chapters 1 and 2,[1] in addition to the mandatory standpoint I am obliged to take up by reason of the temporal and personal context of utterance, I can imaginatively project myself into other times and other men's shoes, and consider how I should view things then or from their position. Many of our concepts, notably that of justice,[2] are constituted by the requirement of invariance under the conjugation of the standpoints we can imaginatively take up as we consider the issue from all points of view. In this chapter we shall be concerned with this latter sort of question; with alteration, not of the mandatory temporal standpoint S each speaker is obliged to occupy because of the temporal context of his utterance, but of the voluntary temporal standpoint R he is free to imagine himself in. Like Prior's tense logic, the logic of temporal standpoints gives only a static picture of the discourse

---

[1] See above, ch. 1, pp. 11–12, ch. 2, § (iii) pp. 20–21.
[2] See J. R. Lucas, *On Justice*, Oxford, 1980, esp. chs. 1 and 3; 'Philosophy and Philosophy of', *Proceedings of the British Academy*, 1986, p. 263.

of one speaker at one time.[3] It is concerned with the RE core of that discourse, and the inferences between different RE propositions, as R, the temporal standpoint, varies, and with the truth to be ascribed to them. Intuitively we accept certain inferences, such as that from 'I was in Cambridge' to 'I have been in Cambridge' and *vice versa*, but are queasy about others, such as that from 'I was in Cambridge' to 'I had been going to be in Cambridge'. The task of this chapter is to formulate rules of inference to legitimise the former but not the latter.

Typically, as we saw in our criticism of Prior's tense logic,[4] tensed statements have an implicit reference to a specific date: 'I was in Cambridge' means not Prior's Pp, 'There *be* some time before now at which I *be* in Cambridge', but 'I was in Cambridge at the time we were speaking about – yesterday/last week/last term etc.', so that the negation is not $H-p$ 'At every time before now I be not in Cambridge' but 'I was not in Cambridge *at the time in question*'. There are, however, two qualifications to be made. First, not all tensed utterances refer to specific times. The perfect tense 'I have been in Cambridge' characterizes the present state of affairs with reference to my having been in Cambridge at some time previously, and though some specific time may be understood, on some occasions it means simply that I have been in Cambridge at some time or other. So equally, 'I have never been to Rome' characterizes my present state with reference to all previous times, which were all my-not-being-in-Rome ones. Hence it is that, as Grosseteste pointed out,[5] sentences of the latter form can be true when uttered at one time and false when uttered subsequently, thereby apparently contradicting the indefectibility of temporal truth;[6] in the same way the sentence 'I have been in Cambridge' can be false when uttered at one time but true on a later

---

[3] See above, ch. 5, § (v), p. 102.

[4] See above ch. 5, § (v), pp. 98–9.

[5] See Grosseteste, *Beiträge*, p. 165, ll. 14–19 (main text) and ll. 31–2 (alternative text); it was rediscovered independently by Kenny, to whom I am indebted for the reference.

[6] See above, ch. 4, § (i), pp. 56-7, § (ii), pp. 63–5.

occasion. But although the truth-value of these sentences changes, it is because of the covert quantifier over all previous dates, contained in the sentence, which ranges over different stretches of time on different occasions of utterance. It is not a counter-example to indefectible truth, which is borne by propositions rather than by sentences, but an instance of the changeable truth characteristically borne by sentences. In the same way the posterior present can be used to refer to an indefinite later date, as in 'We are all going to die', which expresses the common mortality of man. But idiom is variable: if I announce 'I am going to die', I shall be taken to be imparting sad tidings of terminal disease, not a platitudinous pronouncement of my also partaking in the human condition.

Secondly, the temporal reference is often not very precise. We view not only instantaneous events but continuing activities,[7] and we only vaguely indicate when to expect the essay to be finished, the book to be returned, or our dropping in for a cup of tea. Nevertheless, it is reasonable to idealise, and to attach precise specific dates to propositions and propositional contents, recognising that in practice the dates will often be imprecise and sometimes needing to be expressed by means of existential or universal quantifiers.

We can indicate dates by subscripts. There is no difficulty when we are dealing with simple propositional contents, such as there *being* a sea battle; if p stands for there *being* a sea battle, we can represent there *being* a sea battle on February 29th, 2000, by Seabattle$_{\text{February 29th, 2000}}$, or p_{29} for short. More generally we shall use the letters R, s, t, u, v, w, as subscripts indicating dates, with the convention that, unless otherwise indicated, the temporal order is the same as the alphabetical order, except that since we sometimes seek a generality irrespective of order, the subscript R is not necessarily either before or after any other.[8] But we are not dealing only with

---

[7] See above ch. 2, § (iii), p. 22.

[8] It is slightly awkward using R both as a reference point and as a subscript indicating a date. But subscript r is easily confused with subscript t, and logical awkwardness is less bad than typographical confusion.

propositional contents. We need to indicate temporal stand-points. Two symbolisms are available. One, suggested to me by Prior, is to use brackets to enclose the proposition – usually, but not necessarily, a propositional content – and a subscript date. Thus, again suppressing the month and the year, we could symbolize the proposition that on February 28th, 2000, there *be* going to be a sea-battle on February 29th, 2000, by $[p_{29}]_{28}$. This symbolism has the advantage of inviting iteration: we can nest one proposition inside another, and consider whether $[[p_{29}]_{28}]_{29}$ and $[[p_{29}]_1]_{29}$ entail each other or not. Alternatively, as in the previous chapter, we can draw upon the analogy with modal logic,[9] and use the modal operators, $\Box$ or L, and $\Diamond$ or M, or the corresponding G and H, and F and P, with suitable subscripts to flag the relevant dates, thus: we should still symbolize there *being* a sea-battle on February 29th, 2000, by p_{29}, but symbolize the proposition that on February 28th, 2000, there *be* going to be a sea-battle on February 29th, 2000, by $\Box_{28}p_{29}$ or $G_{28}p_{29}$. This has the advantage of being closer to Prior's own notation, and of enabling us to emphasize the distinction between what surely shall happen – $\Box$ or L or G – and what merely may – $\Diamond$ or M or F. Since the future and past are modally different, we need to be able to distinguish future operators from past ones. The subscripts, which indicate the dates, and therefore their temporal order, do this, but we can add emphasis on occasion by having different letters, as Prior does, but with subscripts, thus $G_{28}p_{29}$.[10] In this chapter, where I am concerned primarily with developing a logic, I shall use the bracket notation: in the next chapter and in chapter 8 section (v), where I address myself to more modal questions, I shall use modal operators.

---

[9] See above, ch. 5, § (iv), p. 95.

[10] Alternatively, we could follow the principle of ch. 5, § (iv) pp. 93–5, and use the superscript^{-1} to indicate the inverse of a tense operator; in that case we should write $\Box_{28}p_{29}$, but $\Box^{-1}{}_{28}p_{27}$: but though doctrinally clear, it is typographically awkward.

The propositions of tense logic are like those of ordinary propositional calculus, except that each is dated. It seems reasonable to allow conjunction, disjunction, implication, etc. between propositions of the same date and to disallow it between those of different dates. The restriction of their both having to have the same date goes beyond ordinary usage, where we often have implications across dates – 'If you do not mend the roof, you will get dry rot' – and sometimes conjoin and disjoin differently dated propositions – 'They got married and had lots of children', 'You get out of my orchard, or your father will hear of it'. But if we are citing two propositions in the same logical breath, it seems natural to suppose that we are really doing so from the same temporal standpoint, and reasonable to insist upon it when we are trying to develop a tense logic sensitive to differences of temporal standpoint. We therefore lay down these three formation rules:

1. If C is a well-formed formula, so is $[C]_t$.
2. If C_t is a well-formed formula, with subscript t, then $-C_t$ is also a well-formed formula, with subscript t.
3. If C_t and J_t are both well-formed formulae with the same subscript, then $C_t \rightarrow J_t$, $C_t \& J_t$, and $C_t \vee J_t$ are well-formed formulae too, with the same subscript.

In effect, if we want to combine, by means of a sentential connective, two well-formed formulae, we must first make them of the same date, which we can do by adopting the same temporal standpoint for them both, often, in fact, viewing one from the temporal standpoint of the other; thus, to conjoin, say, C_t and J_u, we must nest one in square brackets with the same subscript as the other, $C_t \& [J_u]_t$ or $[C_t]_u \& J_u$.

These formation rules, though not based on ordinary usage, are defensible. But if we ban *implications* of the form $C_t \rightarrow J_u$ as ill formed, we must allow *inferences* across dates. Although $C_t \rightarrow J_u$ is disallowed,

$$C_t \vdash J_u$$

that is,   from C_t infer J_u,

is under some conditions permitted. Though any utterance of

mine must be uttered all in one logical breath and at one time, it is perfectly permissible to infer from what one person said at one time to something that could be legitimately said by him or someone else at another time. It means that, contrary to the tendency of modern formal logic,[11] our modal logic will sometimes have to be formulated in terms of rules of inference instead of axioms, but since that is what Aristotle did,[12] it can hardly be objected to on fundamental grounds.

The axioms and rules of inference of the minimal modal logic K[13] must hold if the logic of temporal standpoints is to be inferentially transparent, which we certainly expect it to be. That is, we have the Rule of Inference

RT $\quad\quad\quad\quad$ If $\vdash$ C then $\vdash$ [C]$_t$,

that is to say,

$\quad\quad\quad$ if C is a theorem, then so is [C]$_t$,

which secures that a theorem is a theorem from whatever temporal standpoint it is viewed, and the Axiom K

K $\quad\quad\quad$ $[p_R \to q_R]_t \to ([p_R]_t \to [q_R]_t),$

for any t and R (irrespective of temporal order), which secures that implications that hold when viewed from one standpoint hold also between the same propositions themselves viewed from that standpoint.

It is reasonable also to postulate the axiom D

D $\quad\quad\quad$ $[p_R]_t \to -[-p_R]_t$ for any t and R
$\quad\quad\quad\quad\quad\quad$ (irrespective of temporal order)

The axiom D guarantees a certain freedom from modalised contradiction;[14] it rules out the possibility of our having in our modal discourse an expression of the form

$$[p_R]_t \ \& \ [-p_R]_t.$$

[11] See above, ch. 5, § (i), pp. 83–4.
[12] *Prior Analytics*, I, 15, 34ᵃ22–24.
[13] See above, ch. 5, § (i), p. 85.
[14] See above, ch. 5, § (i), pp. 86–7.

It is a reasonable requirement. The only difficulty was that it might fail to hold at the beginning or end of time, if there were such. But if t were the first moment of time, then p_R would be ill-formed for R earlier than t; and similarly if t were the last moment of time, then p_R would be ill-formed for R later than t. The objection raised in the previous chapter[15] to postulating D or its inverse D^{-1} does not apply to a tense logic in which dates are specified. So we lay down D. Once we have made the dates definite, we never can accept 'It will be (was/has been/is going to be) the case that p_s and it will not (was/has been/is going to be) the case that p_s.'

In addition to the rules and axioms of the system D, we need three principle special to tense logic. The most important of these from the formal point of view is that permitting the introduction and elimination of vacuous temporal modal standpoints. This is provided for by

axiom 1 $\vdash p_t \leftrightarrow [p_t]_t.$

It is because this axiom holds, that Reichenbach's reference point escaped the notice of previous logicians, and often seems artificial and pedantic even now.[16] If $p_{29} \leftrightarrow [p_{29}]_{29}$, then to give the traditional analysis of the simple future, S–E, is the same as to insist on the analysis Reichenbach gives, S–R,E; the common core, although formally different in the two cases, is logically equivalent.

The next principle is in the form of a rule of inference, because it licenses inferences across different dates.

R2 $[p_R]_t \vdash [p_R]_u$ provided $t < u.$

This is the substantially most important, as it encapsulates the semantic requirement of the indefectibility of temporal truth, that once a temporal proposition is true, it remains true thereafter.

---

[15] Ch. 5, § (i), p. 89, § (iv), pp. 96–7.
[16] See above, ch. 2, (iii), pp. 21–22.

In an earlier treatment of the logic of dates and tenses[17] I posited a third principle, which could be expressed here as a rule of inference,

R3′ $\qquad [p_t]_v \vdash [p_t]_u$ provided $t < u$.

This rule licensed the inference from the perfect to the aorist; from 'I have been in Cambridge' to 'I was in Cambridge': and similarly from the future perfect to the simple future; from 'I shall have been in Cambridge' there follows 'I shall be in Cambridge'. Propositions from a later temporal standpoint entail propositions from an earlier, so long as the earlier standpoint is still later than that of what is being viewed. Once an event has taken place, it is unalterable, and must be seen in the same light as regards truth and falsity at all subsequent times, earlier as well as later. It is only when we are viewing not retrospectively but prospectively that a shift to an earlier standpoint is not modally innocuous.

The rule is valid, but need not be postulated separately, as it can be derived from the rule **R2** and axiom 1. For by axiom 1

$$\vdash p_t \rightarrow [p_t]_t;$$

Hence, contraposing,

$$\vdash -[p_t]_t \rightarrow -p_t.$$

But also (substituting $-p$ for p)

$$\vdash -p_t \rightarrow [-p_t]_t,$$

so

$$\vdash -[p_t]_t \rightarrow [-p_t]_t.$$

Thus we see that the axiom 1 implies that the operator $\square_t$ when applied contemporaneously commutes with negation and so is modally vacuous. Also, since we can use **R2** to argue from an

[17] J. R. Lucas, *A Treatise on Time and Space*, London, 1973, p. 288.

earlier to a later temporal standpoint, we can argue in a *Modus Tollendo Tollens* fashion from a later to an earlier with respect to negations: that is,

from **R2**

$$[p_t]_t \vdash [p_t]_u$$

we obtain

$$-[p_t]_u \vdash -[p_t]_t.$$

By contemporaneous commutativity, this yields

$$-[p_t]_u \vdash [-p_t]_t,$$

which in turn is subject to **R2**, and yields

$$-[p_t]_u \vdash [-p_t]_u,$$

and

$$-[p_t]_u \vdash [-p_t]_v,$$

for any u,v such that $t \le u \le v$. But in virtue of D

$$[-p_t]_v \vdash -[p_t]_v;$$

hence

$$-[p_t]_u \vdash -[p_t]_v,$$

and, arguing once again in a *Modus Tollendo Tollens* fashion,

$[p_t]_v \vdash [p_t]_u$ provided that $t \le u \le v$.[18]

We have thus established R3′ as a derived rule of inference. The limiting case,

$$[p_t]_v \vdash [p_t]_t.$$

[18] For one apparent counter-example, see Richard Sorabji, *Time, Creation and the Continuum*, London, 1983, who argues on p. 12, 'if something *has* ceased (perfect tense), it does not follow straight off that there is a particular time at which it ceased (aorist)'. But this, as Sorabji admits, is something special about the grammar of the verb 'to cease'.

is equivalent to

$$-[p_t]_t \vdash -[p_t]_v.$$

Since we already have, by virtue of **R2**, $-[p_t]_v \vdash -[p_t]_t$, we can prove

$$-[p_t]_v \vdash [-p_t]_v.$$

This result is of considerable interest. It is the converse of D^{-1}, D^{-1*} as we might call it. It shows that in the *past* tense to say that it was not true that p_t implies that it is true that not-p_t. This implication does not hold in the future. The past is 'modally vacuous',[19] whereas the future is modally live. What we have now seen is that the modal vacuity of the past follows from that of the present together with **R2**, which is the syntactic corollary of the semantic principle of the indefectibility of temporal truth.[20]

Other axiomatizations of the logic of temporal standpoints are possible. We can show a partial analogy with T by replacing axiom 1 by a one-way implication,

axiom 1* $\qquad\qquad \vdash p_t \rightarrow [p_t]_t,$

and having a general T-like rule of inference

T* $\qquad [p_t]_r \vdash p_t$ for all R, after as well as before t.

It might seem that this T-likeness ran counter to our stipulation in chapter 5 that tense logics should not include T.[21] But here we are dealing not with the mandatory shifts of tense as between one occasion of utterance and another, but with the optional shifts of temporal standpoint within the context-independent RE core of tensed discourse. No shift of temporal standpoint could make a proposition true unless its propositional content was itself true. And so a T-like rule of inference would be acceptable, since it does not show that the past or the

---

[19] See above, ch. 5, § (i), p. 86.
[20] See above, ch. 4, § (i), p. 56–7.
[21] See above, ch. 5, § (iv), condition 3, on pp. 94–5.

future imply the present, but licenses inferences from the truth of propositions from a temporal standpoint to the truth of their propositional contents.

Another simplification is available, provided we adopt one further principle, which is desirable on other grounds anyhow. The principle is that we should ascribe to propositions that *be* timelessly or omnitemporally True a particular date – 0 on one reckoning, – ∞ on another[22] – earlier than any other date. We are saying in effect that they were true *ante omnia saecula*, 'from the beginning of time'. Granted this dating convention, if C is a theorem, it has the subscript 0, and can be written J_0. $[J_0]_0$ then follows by axiom 1, and thence by **R2** $[J_0]_t$ for any t. We thus have established **RT** as a derived rule of inference. But we have established more. The convention applies not only to timeless truths, such as theorems of logic, but to omnitemporal ones, such as laws of physics, and it is useful to be able to import all these into temporal discourse without more ado. We thus can offer, as one alternative axiomatization of the logic of temporal standpoints,

Convention 1. 0 is the date of all theorems of logic and other timeless and omnitemporal truths.

axiom 1* $\vdash p_t \rightarrow [p_t]_t,$
T* $[p_t]_s \vdash p_t$ for all *s*, after as well as before *t*,
R2 $[p_R]_t \vdash [p_R]_u$ provided $t < u$.

It is clear that in either axiomatization the system is syntactically consistent, since if we interpret the operator $[\ ]_t$ as a null operator, we are left with propositional calculus. It is not syntactically complete, since the converse of D is not a theorem, but could be added to the system without thereby making it inconsistent. That is to say

D* $-[p_u]_t \rightarrow [-p_u]_t$

---

[22] The symbol ∞ stands for infinity. Since it is easy to transform, *e.g.* by an exponential function, a scale from ∞ to +∞ into one from 0 to +∞, 0 will be used in this passage for convenience' sake.

does not in general hold for $t < u$, but if it were added as an axiom, it would have the effect only of making $[\ ]_t$ modally vacuous, which would reduce the system to propositional calculus, but not make it inconsistent. Since D^* does not hold for $[\ ]_t$, whereas the mirror image D^{-1*} does hold for $[\ ]^{-1}{}_t$, the system does not have the mirror image property, and so B is not a theorem. Nor is T, though the system is T-like. The analogue to **4** is in the $\Box$ terminology

$$\vdash \Box_t p_v \rightarrow \Box_u \Box_t p_v,$$

which is equivalent to

$$\vdash [p_v]_t \rightarrow [[p_v]_u]_t.$$

This is a theorem. For

$$\vdash [p_v]_t \rightarrow [p_v]_u \text{ by } \mathbf{R2}; \text{ whence}$$

$$\vdash [[p_v]_t \rightarrow [p_v]_u]_t$$

and

$$\vdash [[p_v]_t]_t \rightarrow [[p_v]_u]_t \text{ by } \kappa;$$

this together with axiom 1 yields the result

4' $\qquad \vdash [p_v]_t \rightarrow [[p_v]_u]_t.$

Thus the logic of temporal standpoints satisfies the conditions for being a tense logic.

(ii) Semantics

The propositional core of tensed discourse is independent of occasion of utterance, and so is suited to be a bearer of some omnitemporal or timeless truth. If I can truthfully say 'By February 28th I shall have been in Cambridge', on February 28th I shall be able truthfully to say 'I have been in Cambridge', and on February 29th I shall be able truthfully to say 'On February 28th I had been in Cambridge'. It is reasonable therefore to ascribe the truth-value TRUE to the proposition that views my having been in Cambridge from the standpoint of February 28th; or to say that the proposition *be* True. Mathematically we can construe this as a function from an ordered pair of a date together with a proposition (usually, as in this case, a propositional content) into the set {TRUE, FALSE}. But the same mathematical function from an ordered pair of a date together with a proposition into the set {TRUE, FALSE} can be viewed in a different light. Instead of construing the date as part of the proposition, the temporal standpoint, corresponding to Reichenbach's reference point R, we can take it as qualifying the TRUE or FALSE, converting them into temporal truth and temporal falsity, ascribed at the time stated. Instead of understanding it as saying that my having on February 14th been in Cambridge on February 12th *be* True, we take it as saying that my being in Cambridge will be/is/was temporally true on February 14th. The formal structure is the same: the interpretation differs in imputing the temporality in one case to the proposition and in the other to the truth.

The ascription of truth is not straightforward. Formally, we have a function from an ordered pair into {TRUE, FALSE}, and even if we tailor propositional contents so that the Principle of Bivalence holds of them and they *be* always either True or False, we have no warrant for supposing that the function from an ordered pair is likewise a total function. If we consider temporal truth first, we find that it does not satisfy the Principle of Bivalence in our ordinary way of thinking. We speak of statements about the future coming true, but we do not thereby mean that previously they were false. If once a prediction is falsified, it remains false for ever after. Temporal falsity and temporal truth are alike indefectible. Once acquired, they cannot be lost, but it is possible to acquire them. Hence the function representing temporal truth and falsity cannot be a total function into the two-membered set {TRUE, FALSE}. Whether we should think of it as a partial function into the two-membered set {TRUE, FALSE}, or as a total function into the three-membered set {TRUE, NEITHER, FALSE} is not clear;

ordinary usage gives no guidance. All we are sure of is that at one time hopes and fears are neither true nor false, and at a later time are either fulfilled or disappointed or dispelled. They are, as it were, in a limbo until the moment of truth. Whether that limbo is a definite place with a position of its own on the logical map, or simply a non-place, is unclear. Most naturally we think of it as a non-place, that is to say we construe the function as a partial function into {TRUE, FALSE}, but under pressure we are inclined to concede that it must be a place and postulate an extra truth-value NEITHER, which is permissible, so long as we do not seek then to give a truth-functional account of the sentential connectives.[23]

A similar account can be given of the ascription of truth to RE propositions. Because these express the context-independent core of tensed discourse, the truth that they can bear is a timeless truth. What varies with the temporal standpoint is not temporal truth but the propositions themselves, some being definitely True or False, others not. With a contemporaneous or subsequent standpoint the RE proposition thereby formed *be* either True or else False; but this is not necessarily so with an earlier one. Instead of not being able to say that a propositional content (Reichenbach's E) is (temporally) either true or false before the event, we are unable to say that the complex RE proposition expressed by the corresponding prediction *be* (timelessly) either True or False. And then again we have the choice of saying either that it *be* neither True nor False, or else that it *be* Neither.

We can insist on the Principle of Bivalence, but at a price. In the one case, as we have seen,[24] we import a modal force to 'true' and 'false' in their temporal applications. In the other, as we are now able to see, the modal force is carried by the temporal operators. If we insist that $[p_u]_t$ *be* in all cases either True or else False, then $[\ ]_t$ becomes a full-blooded, non-vacuous operator for which the thesis D* does not hold. By insisting on

---

[23] See above ch. 4, § (iv), pp. 74–5, and ch. 5, § (ii), 90–92.
[24] See above ch. 4, § (iv) pp. 74–75, 77–8.

the Principle of Bivalence for the posterior present tense 'is going to', we make it mean 'must' – or else perhaps 'perhaps'. If we stipulate that it *be* True only when the partial function assigns the truth-value TRUE, and otherwise False, then we are giving 'is going to' the sense of 'surely shall', but allowing that it could *be* False that there is going to be a sea battle tomorrow, even though when tomorrow comes there is one, because at present (to move into the terminology of temporal truth) it is not definitely true – the admiral has not made up his mind. Alternatively, unidiomatically but conceivably, we could stipulate that 'is going to' *be* False only when the partial function assigns the truth-value FALSE, and otherwise True, then we would be giving 'is going to' the sense of 'may be' or 'perhaps'. In the former construal both 'There surely shall be a sea battle' and 'There surely shall not be a sea battle' can both *be* False: in the latter 'There may perhaps be a sea battle' and 'There may perhaps not be a sea battle' can both *be* True. Our idiom 'is going to' is closer to the former than the latter, but is not clearly subject to the Principle of Bivalence, and we often hesitate to ascribe a truth-value to predictions before the event.

7

Contingency Planning for Naval Logicians

(i) Syntactic Approach

Four distinctions are needed in order to unravel the two arguments that puzzled Aristotle in chapter 9 of *De Interpretatione*. The first is Reichenbach's reference point, R, indicating the temporal standpoint we choose to take up, as opposed to S, the date of utterance, and E, the date of the event, neither of which are for us to choose. The second distinction, which follows from this, is that between the simple future, of the form S–R,E, and the posterior present, of the form S,R–E. The third is between different sorts of truth and, concomitantly, different bearers of truth. And the fourth is between the different modalities involved in predicting; that is to say the different modalities with respect to which a future contingent is contingent.

Aristotle has two puzzles: the 'white now' argument, which is concerned with the different tenses which have to be used in order to say the same thing at different times, and tomorrow's sea battle, which is concerned with the logical grammar of the future tense and of necessity with respect to negation and disjunction. Aristotle does not distinguish, as modern logicians do, between the syntactic and the semantic approach, the former being entirely concerned with logical grammar, the latter with interpretations and, in particular, with truth. It is helpful to schematize the discussion of Aristotle according to the modern distinction, though recognising that Aristotle's terminology does not fit, and may be misrepresented by the account given.

Aristotle argues first:

ἔτι εἰ ἔστι λευκὸν νῦν, ἀληθὲς ἦν εἰπεῖν πρότερον ὅτι ἔσται λευκόν,
ὥστε ἀεὶ   ἀληθὲς ἦν εἰπεῖν ὁτιοῦν τῶν γενομένων ὅτι ἔσται· εἰ δ'
ἀεὶ ἀληθὲς ἦν εἰπεῖν ὅτι ἔστιν ἢ ἔσται, οὐχ οἷόν τε τοῦτο μὴ εἶναι
οὐδὲ μὴ ἔσεσθαι. ὃ δὲ μὴ οἷόν τε μὴ γενέσθαι, ἀδύνατον μὴ γενέσθαι·
ὃ δὲ ἀδύνατον μὴ γενέσθαι, ἀνάγκη γενέσθαι· ἅπαντα οὖν τὰ ἐσόμενα
ἀναγκαῖον γενέσθαι.

If something is white now, it was correct (Greek ἀληθὲς [*alethes*],
true) to say beforehand that (Greek ὅτι [*hoti*], literally that, but often
used in lieu of quotation marks) it will be white; and thus of a past
event it was always correct (Greek ἀληθὲς [*alethes*], true) to say
(Greek ὅτι [*hoti*], literally that, but often used in lieu of quotation
marks) it will be. If it was always correct (Greek ἀληθὲς [*alethes*],
true) to say (Greek ὅτι [*hoti*] literally that, but often used in lieu of
quotation marks) it is or it will be, it could not be that it neither was,
nor was going to be, the case. It is impossible that something should
not happen which is such that it could not be that it would not
happen. If it is impossible for it not to happen, it is necessary that it
should happen. With all future events, therefore, it is necessary that
they should happen.[1]

If something is white now – the tense is clearly present, S,R,E,
with the time of utterance, the reference point and the date of
the event all being contemporaneous. What then was it correct
(Greek ἀληθὲς [*alethes*], true) to say beforehand? The RE part
was the same, but the time of utterance was at that time earlier
than the reference point and the date of the event, and so would
be of the form S–R,E, that is, the future simple. If someone had
uttered the words 'It will be white', he would have spoken
correctly. He would have been guilty of an inconsistency if he
had denied the earlier utterance, saying 'It will not be white',
and then affirmed the latter, saying 'It is white'. The present
tense uttered contemporaneously and the future simple uttered
at an earlier time express the same RE proposition, and are
logically equivalent. But when we move into indirect speech, we

[1] *De Interpretatione*, 18^{b}9–15, based on the translation of J. L. Ackrill,
Aristotle's Categories and De Interpretatione, Oxford, 1966.

introduce a further reference point contemporaneous not with the event but with the time of the putative utterance. Instead of saying in direct speech 'It will be white', we should report that it was true to say that it would be white. Besides the possibly redundant reference point contemporaneous with the event we are required by our standard rule for indirect speech to phrase what is reported from the temporal standpoint of original speaker. Logically speaking, we should say looking back on its being white,

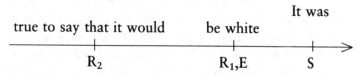

Greek had no way of writing inverted commas. Aristotle uses ὅτι [*hoti*] which also means 'that', and is the standard way of introducing an indirect statement. He uses the future simple, saying ὅτι ἔσται [*hoti estai*], but the logical pressure to conform to the rules for indirect speech would be at work in Greek as well as in English, and without an idiomatic posterior present tense being available in either language, there would be little to resist the transition, especially when the infinitive, ἔσεσθαι [*esesthai*] and participle ἐσόμενα [*esomena*] are being used. And therefore, instead of the pedantically precise construal above, we understand *it was true to say that it would be white* thus:

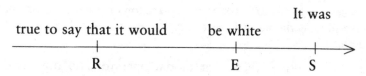

which we could express by the retrospective statement that *it was going to be* or, making the modal force more explicit, *it was to be*. The rules for indirect speech have introduced an additional reference point, which in view of the apparent redundancy of the original reference point, contemporaneous with the event, has become the only reference point, antecedent to the event, and therefore ineliminable and modally live. If we consistently use direct speech and distinguish the posterior present from the future simple, no problem arises. If we use indirect speech, we can avoid the fallacy by distinguishing the original reference point R_1 from that of the ἀληθὲς ἦν εἰπεῖν [*alethes en eipein*], it was true to say, the former indicating the temporal standpoint of the invariant RE proposition we can truthfully affirm, if indeed this is white now, while the latter, R_2, indicates only the date of hypothetical utterance as described after the event, and is not an actual temporal standpoint at all, but merely a way of referring to a hypothesized S that never actually took place. Once the distinction between R_1 and R_2 is drawn, the temptation to assimilate them disappears, and we no longer feel impelled to slide from the innocuous R_1 to the loaded R_2.

Some of the problems of tomorrow's sea battle are similarly syntactic.

καὶ ἐπὶ τῆς ἀντιφάσεως ὁ αὐτὸς λόγος· εἶναι μὲν ἢ μὴ εἶναι ἅπαν ἀνάγκη, καὶ ἔσεσθαί γε ἢ μή· οὐ μέντοι διελόντα γε εἰπεῖν θάτερον ἀναγκαῖον. λέγω δὲ οἷον ἀνάγκη μὲν ἔσεσθαι ναυμαχίαν αὔριον ἢ μὴ ἔσεσθαι, οὐ μέντοι γενέσθαι αὔριον ναυμαχίαν ἀναγκαῖον οὐδὲ μὴ γενέσθαι· γενέσθαι μέντοι ἢ μὴ γενέσθαι ἀναγκαῖον.

And the same account holds for contradictories: everything necessarily is or is not, and will be or will not be; but one cannot divide and say that one or the other is necessary. I mean for example: it is necessary for there to be or not to be a sea battle tomorrow; but it is not necessary for a sea battle to take place tomorrow nor for one not to take place – though it is necessary for one to take place or not to take place.[2]

Aristotle's point here is that modal operators do not divide under disjunction, or – what comes to the same thing – commute with negation.[3] In general

$$\Box(p \lor q) \text{ does not entail } \Box p \lor \Box q.$$

---

[2] *De Interpretatione*, 19ª27–32.
[3] See above, ch. 5, § (i), p. 89.

We have distinguished two modalities: temporal necessity L_t[4] and in the posterior present tense the RE operator $[p_u]_t$ where the date t is before that of the proposition p_u. In each case

$$L_t(p_u \vee -p_u) \not\vdash L_t p_u \vee L_t - p_u, \text{ that is,}$$

necessary$_{now}$(seabattle$_{tomorrow}$ or noseabattle$_{tomorrow}$)

does **not** entail

(necessary$_{now}$seabattle$_{tomorrow}$) or
(necessary$_{now}$noseabattle$_{tomorrow}$);

and $\qquad [p_u \vee -p_u]_t \not\vdash [p_u]_t \vee [-p_u]_t$, that is,

there is$_{now}$ going to be (seabattle$_{tomorrow}$ or noseabattle$_{tomorrow}$)

does **not** entail

(there is$_{now}$ going to be seabattle$_{tomorrow}$) or
(there is$_{now}$ going to be noseabattle$_{tomorrow}$).

These two modalities are very similar: they differ in that the former satisfies the Principle of Bivalence and has no truth-gaps, while the latter, as we have defined it, has truth gaps, and sometimes, when t is before u, is neither True nor False. There is a further difficulty that the English phrase 'is going to', together with the ordinary undifferentiated future tense in most European languages, commutes with negation: 'there is not going to be a sea battle tomorrow' is taken to mean 'there is going to be no sea battle tomorrow', though grammatically it looks like 'it is not the case that there is going to be a sea battle tomorrow'. There is thus another line of argument pressing itself on us:

either   (there is$_{now}$ going to be seabattle$_{tomorrow}$)
or   (there is$_{now}$ not going to be seabattle$_{tomorrow}$),

---

[4] Note that L is no longer being reserved for logical necessity, as it was in ch. 3, section (iv), pp. 41–3. Although we could use □, it is helpful in this chapter to make a clear typographical distinction between the bracket notation of the RE calculus and ordinary modal operators.

which is a tautology, an evident instance of the Law of the Excluded Middle, and seems to yield

(there is$_{now}$ going to be seabattle$_{tomorrow}$) or
 (there is$_{now}$ going not to be seabattletomorrow),

or equivalently,

(there is$_{now}$ going to be seabattle$_{tomorrow}$) or
 (there is$_{now}$ going to be noseabattle$_{tomorrow}$).

Temporal necessity is not just confused with the, itself confusing, posterior present tense. The modal properties of temporal necessity themselves vary with the temporal standpoint taken up. Whereas it is in general false to say

$$\vdash L_t p_u \vee L_t - p_u$$

(necessary$_{now}$seabattle$_{tomorrow}$) or
 (necessary$_{now}$noseabattle$_{tomorrow}$),

it *is* a theorem when t is contemporaneous with or after u: thus

$$\vdash L_u p_u \vee L_u - p_u ,$$

that is,

(necessary$_{tomorrow}$seabattle$_{tomorrow}$) or
 (necessary$_{tomorrow}$noseabattle$_{tomorrow}$).
Tomorrow when it comes must be a sea-battle day or a non-seabattle day, and that then and thereafter will be a divided necessity.

Τὸ μὲν οὖν εἶναι τὸ ὄν, ὅταν ᾖ, καὶ τὸ μὴ ὄν μὴ εἶναι, ὅταν μὴ ᾖ, ἀνάγκη.

(That which is, necessarily is) when it is: and (that which is not, necessarily is not) when it is not.[5]

The standard condition for non-vacuity does not apply for present necessity, and hence not for past necessity either. Necessity can sometimes be divided, and whether it can or

---

[5] *De Interpretatione*, 19ª273–24; my bracketing.

cannot depends on the temporal standpoint taken up, itself a matter of arbitrary choice. From the point of view of the future the future is necessary, and what will be must be.[6] But future necessity is not present necessity, and

$$L_u p_u \text{ does not entail } L_t p_u \text{ that is,}$$
necessary_{tomorrow}seabattle_{tomorrow}

does not entail necessary_{now}seabattle_{tomorrow}.

Once the point is made that temporal necessity depends upon the temporal standpoint taken up, the fallacious inference is easily seen to be fallacious. But its power to puzzle is equally easy to understand.

(ii) Timeless Truth

These resolutions are entirely syntactic, in terms of a confusion of reference points, an illegitimate shift from a later to an earlier one, and the modal properties of tense logic and temporal necessity. Aristotle, however, poses the problem in terms of the semantic concept of truth.

εἰ ἔστι λευκὸν νῦν, ἀληθὲς ἦν εἰπεῖν πρότερον ὅτι ἔσται λευκόν, ὥστε ἀεὶ ἀληθὲς ἦν εἰπεῖν ὁτιοῦν τῶν γενομένων ὅτι ἔσται· εἰ δ' ἀεὶ ἀληθὲς ἦν εἰπεῖν ὅτι ἔστιν ἢ ἔσται, οὐχ οἷόν τε τοῦτο μὴ εἶναι οὐδὲ μὴ ἔσεσθαι.

If something is white now, it was true to say beforehand 'It will be white'; so that it was always true to say of anything that has happened that 'It will be so'. But if it was always true to say that 'It is so' or 'It will be so', it could not not be so, or not be going to be so.[7]

and

οὐδὲν γὰρ κωλύει εἰς μυριοστὸν ἔτος τὸν μὲν φάναι τοῦτ' ἔσεσθαι τὸν δὲ μὴ φάναι, ὥστε ἐξ ἀνάγκης ἔσεσθαι ὁπότερον αὐτῶν ἀληθὲς ἦν εἰπεῖν τότε. ἀλλὰ μὴν οὐδὲ τοῦτο διαφέρει, εἴ τινες εἶπον τὴν

[6] See below, ch. 8, § (iii), pp. 145–6; (v), pp. 152–4. See p.120 for more careful translation.
[7] *De Interpretatione*, 18^b9–13.

ἀντίφασιν ἢ μὴ εἶπον· δῆλον γὰρ ὅτι οὕτως ἔχει τὰ πράγματα, κἂν
μὴ ὁ μὲν καταφήσῃ ὁ δὲ ἀποφήσῃ· οὐ γὰρ διὰ τὸ καταφάναι ἢ
ἀποφάναι ἔσται ἢ οὐκ ἔσται, οὐδ' εἰς μυριοστὸν ἔτος μᾶλλον ἢ ἐν
ὁποσῳοῦν χρόνῳ. ὥστ' εἰ ἐν ἅπαντι τῷ χρόνῳ οὕτως εἶχεν ὥστε τὸ
ἕτερον ἀληθεύεσθαι, ἀναγκαῖον ἦν τοῦτο γενέσθαι, καὶ ἕκαστον τῶν
γενομένων ἀεὶ οὕτως ἔχειν ὥστε ἐξ ἀνάγκης γενέσθαι· ὅ τε γὰρ
ἀληθῶς εἶπέ τις ὅτι ἔσται, οὐχ οἷόν τε μὴ γενέσθαι· καὶ τὸ γενόμενον
ἀληθὲς ἦν εἰπεῖν ἀεὶ ὅτι ἔσται.

For there is nothing to prevent someone's having said ten thousand
years beforehand that this would be the case, and another's having
denied it; so whichever of the two was true to say then, will be the case
of necessity. Nor, of course does it make any difference whether any
people actually made contradictory statements or not. For clearly this
is how the actual things are even if someone did not affirm it and
another deny it. For it is not because of the affirming or denying that it
will be or not be the case; nor is it a question of ten thousand years
beforehand rather than any other time. Hence, if in the whole of time
the state of things was such that one or the other was true, it was
necessary for this to happen, and for the state of things always to be
such that everything that happens happens of necessity. For what
anyone has truly said would be the case cannot not happen; and of
what happens it was always true to say that it would be the case.[8]

Here it is antecedent truth that carries the weight of argument.
So we have to ask: What sort of truth? What is the bearer of
that sort of truth?

We might first of all think of timeless truth, which is
characteristically borne by propositions and propositional
contents. But timeless truth is timeless, and can play no part in
tensed discourse. If it *be* True that there *be* a sea battle on
February 29th, 2000, we cannot meaningfully say that it was
True. To say that a proposition was True would be rather like
saying two and two used to be four, or are going to be next
week. The standard formulation of the fallacious argument thus
falls at the first hurdle: the temporal necessity we need to
attribute to antecedent truth cannot be ascribed to the timeless

---

[8] *De Interpretatione*, 18ᵇ33–19ᵃ6.

truth that the Principle of Bivalence requires us to ascribe to one
of every pair of contradictory propositions.

But that does not conclude the argument. We might still feel
that there was a certain propositional content, that there *be* a
sea battle on February 29th, 2000, which must be either True or
False: if it *be* True, then the utterance 'There will be a sea battle
on February 29th, 2000,' is temporally true; and if it *be* False,
then the utterance 'There will be a sea battle on February 29th,
2000,' is temporally false, so either way there is an antecedent
truth or falsity which is unalterable, and does necessitate the sea
battle's happening or not happening.

Certainly, in the world of abstract entities we can tailor
propositions and *being* True so that the Principle of Bivalence
holds, and every proposition *be* either True or else False; but
such a platonic realm is open to Aristotle's criticism that it is
totally separate from this world, and cannot have any bearing
on it. The Principle of Bivalence assures us that that there *be* a
sea battle on February 29th, 2000 *be* either True or else False,
and the Semantic Law of the Excluded Middle[9] assures us that
either that there *be* a sea battle on February 29th, 2000 *be* True,
or else that there *be* not a sea battle on February 28th, 2000 *be*
True, but neither of these gives us any grip on *which*. We can
say in general terms that every proposition *have* a truth-value,
either TRUE or FALSE, but this does not determine **which**
propositions *have* the truth-value TRUE and **which** FALSE, nor of
any particular proposition which of the two truth-values it
have.[10] We know that every triangle *be* either obtuse, or
right-angled, or acute, but that does not imply that a triangle
have one of these properties, unless it is sufficiently particularised as being a triangle with that property. The utterance
'There will be a sea battle on February 29th, 2000' does not
sufficiently particularise the proposition that there *be* a sea

[9] See above, ch. 4, § (iv), p. 74.
[10] Compare Boethius, *Ad Aristotle de interpretatione*, ed. Meiser, p. 125,
'manifestum esse non necesse omnes adfirmationes et negationes *definite* veras
esse . . .', quoted Jan Lukasiewicz, *Selected Works*, ed. L.Borkowski, North
Holland, Amsterdam, 1970, p. 176.

battle on February 29th, 2000, any more than characterizing a triangle as being isosceles determines it as being obtuse or right-angled or acute. We are easily tempted to suppose the truth is in some way a temporal truth, at present hidden from our eyes, but being made manifest in the event. But timeless truth is not temporal truth, and cannot give rise to any temporal necessity, and abstract entities by themselves are too abstract to have any purchase on real things or the course of actual events.[11]

But, it may be protested, abstract entities are not so abstract that they cannot bear temporal truth: we have constructed a sort of temporal truth that propositions and propositional contents can bear.[12] That is so. But artifice was required in constructing indefectible truth as a function from ordered pairs of dates and propositions into the set {TRUE, FALSE}, and we had to compromise either the Principle of Bivalence or omnitemporality, and whichever we do, we cannot infer earlier temporal truth from later temporal truth. If it is, as of now, February 28th, 2000, temporally true that this *be* white on February 28th, 2000, it does not follow that it was temporally true ten thousand years ago that this *be* white on February 28th, 2000. The temporal truth we have constructed to be borne by propositions does not admit of that inference.

(iii) Guestimates

Aristotle was not concerned with a function from ordered pairs of dates and propositions into the set {TRUE, FALSE}. He was concerned with utterances, actual or hypothetical:

εἰ γὰρ ὁ μὲν φήσει ἔσεσθαί τι ὁ δὲ μὴ φήσει τὸ αὐτὸ τοῦτο,

---

[11] See further, Gilbert Ryle, *Dilemmas*, Cambridge, 1954, ch. 2, 'It was to Be', pp. 16–27. See also Storrs McCall, 'A Dynamic Model of Temporal Becoming', *Analysis*, 44, 1984, p. 176: 'The notion of truth, so to speak, bakes no bread, it simply floats on top of whatever events occur or will occur and in no way constrains or affects the possibility of any of them occurring.'

[12] In ch. 6, § (ii), pp. 116–17.

For if someone says that something will be and another denies this same thing. . .[13]

and

ἀλλὰ μὴν οὐδὲ τοῦτο διαφέρει, εἴ τινες εἶπον τὴν ἀντίφασιν ἢ μὴ εἶπον· δῆλον γὰρ ὅτι οὕτως ἔχει τὰ πράγματα, κἂν μὴ ὁ μὲν καταφήσῃ ὁ δὲ ἀποφήσῃ· οὐ γὰρ διὰ τὸ καταφάναι ἢ ἀποφάναι ἔσται ἢ οὐκ ἔσται,

Nor, of course, does it make any difference whether any people made the contradictory statements or not. For clearly this is how things are even if someone did not affirm it and another deny it. For it is not because of the affirming or denying that it will be or will not be the case, . . .[14]

Ryle questions the latter argument.[15] There is a difference between actual predictions and merely possible predictions. A might-have-been prediction has the same dubious status as a might-have-been bullet: we cannot say that a might-have-been bullet has hit an actual target, and similarly it imports an illegitimate determinateness to speak of a might-have-been prediction actually coming true. In talking of possible as well as actual utterances we are moving from the hard actuality of what was really said by a real person to the shadowy world of propositions which are difficult to pin down, and dangerously apt to slip from our grasp altogether.

Nevertheless there are actual utterances, which are dateable events and clearly suited to bear temporal truth. What sort of utterance might have been made ten thousand years ago which could have been said then to be true? So far as Aristotle's 'It is white now' argument goes, it must construe ἀληθὲς ἦν εἰπεῖν [alethes en eipein], it was true to say, as ascribing valedictory truth, or there would be no good reason for saying that we must ascribe it. And valedictory truth poses no threat to freedom, being only apparently antecedent, and in reality post-dated to

---

[13] *De Interpretatione*, 18ᵃ35–6.
[14] *De Interpretatione*, 18ᵇ36–9.
[15] Ryle, *Dilemmas*, ch. 2, 'It was to Be', pp. 17–20.

after the event. But what if it were at least in part a prediction? Someone speaking ten thousand years ago might conceivably have predicted that it was going to be white now; he might, conceivably, have known geology and meteorology much better than we do, and have had good grounds for believing that there was going to be another ice age. If challenged, he could have given his grounds, and if they were good ones, his prediction would have been true, and we should now be suffering a snowy existence as a result of processes we were unable to control. But this would be a different exercise from merely uttering the words 'it will be white on February 28th, 2000'. In the one case it is a prediction for which grounds can be demanded and must, at least to some extent, be given: in the other it is only an idle guess, conjecture, or bet, for which no grounds can be demanded or need be given.

Although I have drawn a sharp distinction between conjectures and predictions, no such sharp distinction is given us in ordinary language. We have only one future tense, which is ambiguous as between the future simple, S–R,E, and the posterior present, S,R–E. We cannot on linguistic grounds divide utterances in the future tense into conjectures bearing valedictory truth and predictions bearing defeasible truth, nor often can we tell from the context. Indeed, people are sometimes at pains to blur the distinction, as when they speak of guestimates, disclaiming good grounds to warrant their assertions, while not quite conceding that they are mere guesses.

The blurring of the distinction between conjectures and predictions is not just due to linguistic inadequacy, but reflects there being different modalities involved. Often we make guarded predictions on the basis of a weak modality. We say that, granted everyone acts morally, reasonably, or sensibly, some particular event should take place; but those conditions are ones we should be unwise to rely on unreservedly. If the admiral has made up his mind, then a sea battle is going to ensue in the normal course of events, but not if there is a hurricane, or a sudden order to return back to Athens. Even our hardest modalities, those that underlie predictions made in the physical sciences, are subject to some escape clause – 'unless a black hole swallows us all up'. These mixed modalities make it hard to classify utterances in the future tense, and hard also to think clearly about their import. We need to ask not only 'Future contingent *what*?',[16] but 'Contingent in respect of *what modality*?'. In one mode of discourse we can predict: in another all things are possible, and no prediction can be made. Thus it is intelligible to say 'There has not got to be a sea battle – the admiral has not made up his mind yet, and anyway there might be a storm. Nevertheless, I predict that there will be one; mark my words.' On the one hand it disclaims the 'present causes' that would put it into the posterior present tense and make it a genuine prediction; with respect to that modality it is contingent: on the other hand it is more than an idle guess; if there is no sea battle, his words will have to be eaten. We are unclear which mode to adopt. We are torn. On the one hand predictive truth offers guidance with an unconditional warranty, and so we must not claim it lightly or unadvisedly, and must take all factors into account before presuming to issue a prediction: on the other hand guidance is needed, and to play safe, and not to offer any at all is unhelpful. So we say something, which at one level is a prediction, at another a mere conjecture, and are correspondingly unsure what sort of truth it can aspire to.

Some mixture of valedictory truth, borne by conjectures, and defeasible truth, borne by predictions, could seem to threaten freedom. For defeasible truth, if not defeated, is indefectible truth, and the antecedent appearance of valedictory truth confers an air of unalterability on the not-yet-defeated indefectible truth of a fully explicit prediction. It genuinely *was* valedictorily true to say ten thousand years ago that Elizabeth II would be our Queen on February 28th, 1988, because, although any statement to that effect had not then come true, it since has, and we ascribe to the utterance a covertly post-dated truth. It was *not* indefectibly true to say ten thousand years ago that Elizabeth II would be our Queen on February 28th, 1988,

---

[16] See above, ch. 4, § (iii), p. 65.

because it had not come true then. The claim was only defeasible then, vulnerable to subsequent falsification. It is no longer vulnerable now, and is indeed indefectibly true. But whereas with a pure prediction we can resist the temptation to antedate indefectibility, and with a pure conjecture recognise that the antedating is apparent only, it is easy in the mixed case to antedate a truth that is not covertly post-dated, and to suppose that there really was something in the nature of things then that necessitated the events that have happened since. It is only when we have developed somewhat sophisticated semantics for the temporal truth of posterior present propositions in chapter 8, and of genuine future simple propositions in chapter 9, that we can avoid muddling the dates and the modalities of temporal truth.

(iv) Argument

The word 'true' is not used only to vouch for the reliability of predictions or to lay bets or hazard idle conjectures. Often we conjecture in a serious mood as part of serious argument. Many of the pressures on the concept of truth arise from its use as a logical counter, and we can best appreciate their bearing on the logic of future truth by considering the ways we need to argue about the future. Naval staff officers have to make plans. While the admiral is making up his mind, they have to prepare to put his decision into effect, whatever it may be. *Either there is going to be a sea battle tomorrow* – in that case we must grease the rowlocks and generally get ready our ships: *or there is is going to be no sea battle* – in that case we must lay on some games to keep the men's morale up and pre-empt any sulks on the part of some latter-day Achilles. The staff are preparing for a decision in the immediate future which will effectively determine the course of events. But they need also to look ahead, and make contingency plans to prepare for eventualities in the more distant future. Suppose it is true that there is a sea battle tomorrow; then either we shall win, or we shall lose; if we win, we must follow it up with a raid on Naupactus, while if we lose we must take steps to prevent the Locrians wavering in their allegiance. *Suppose it is true that there is a sea battle tomorrow* – we need to be able to address hypothetical situations, and consider what it would be like if they were true. The word 'true' in this use carries no warranty at all: it is simply a logical counter, an essential weapon in the gladiatorial passage of arms, which enables us to explore possibilities and determine what must and must not be the case. Without affirming that it is true, we need to entertain the supposition that it is, and without affirming that it is false, we need to entertain the supposition that it is. Not only do we need to entertain various suppositions, but we need to be able sometimes to discharge them, arguing by dilemma, and in particular from the Law of the Excluded Middle. If we win, we must follow it up with a raid on Naupactus, while if we lose we must take steps to prevent the Locrians wavering in their allegiance; in order to carry out a raid on Naupactus, we need a landing party; in order to terrorise the Locrians, we need a landing party; so, if there is a sea battle tomorrow, we shall need a landing party. Equally, if there is no sea battle tomorrow, in order to keep up morale, we must lay on, besides the games, some serious military exercises, of which preparing a landing party must be one; so, if there is no sea battle tomorrow, we shall need a landing party. But either there will be a sea battle tomorrow or there will be no sea battle tomorrow, hence either way we shall need a landing party. We are looking at the sea battle not from today's standpoint, when, as of now, the admiral has not yet made up his mind whether to engage or stay in harbour, and it is not yet true that there is going to be a sea battle tomorrow nor that there is going to be no sea battle tomorrow, but from tomorrow's standpoint when either there is a sea battle or there is not. The English conditional idiom 'if there is a sea battle', where the present tense contrasts with most European languages, which use the future, may be trying to express the thought that the standpoint envisaged by the antecedent is contemporaneous with the supposed event. The truth borne by suppositions in argument is not the warranted assertibility of the law courts but the free-ranging *fiat* of the speculative logician. Instead of being anchored in the here and now of the responsible utterance of predictions meant to be taken seriously as a guide to action, it takes up the temporal standpoint of the supposed event so as to enable the argument to go on from there.

We can see now why the future simple has a different logic from the posterior present. The important fact about the future simple from a logical point of view is not that the date of the event, E, is after the date of speaking, S, but that it is contemporaneous with the reference point, R. And if we view events from a contemporaneous standpoint, there is no veil of modality between them and us: we are alongside them in our mind's eye, and must envisage their being true or else their being false. *Tertium non datur*, because the limbo into which we put the temporal truth of future propositions has been supposed away. The Principle of Bivalence is a necessity of dialectical argument when two people are arguing with each other, and are trying to explore the consequences of various positions, and because the Principle of Bivalence holds for argument, the Law of the Excluded Middle holds for the future simple, though not for the posterior present.

Thus far we have drawn many distinctions between different types of truth and different bearers of truth, and have formulated some rules for the logic of temporal standpoints which accord with our customary usages. It remains to give a more coherent account of temporal truth, and to offer a semantics for the logic of temporal standpoints.

8

Tree Semantics

(i) Tree Shapes

Tree Semantics, is a development of work by Thomason, McCall, McArthur, McKim and Davis, and Burgess,[1] and is intended to give sense to a proposition's or a propositional content's being said to be temporally true at a time, to provide a semantics for the logic of temporal standpoints, and to distinguish more sharply than we have been able to hitherto between the different senses in which statements about the future can be said to be true.

The 'Tree at time t' is most easily pictured as a set of possible worlds each of which is either accessible from the actual world obtaining at time t, or, conversely, such that the actual world is accessible from it. The accessibility relation, Q, is a 'quasi-ordering', that is to say, it is transitive, reflexive, and anti-symmetrical. It follows that its converse, Q^{-1} must be transitive, reflexive, and antisymmetrical likewise, but the 'mirror image property' does not hold with respect to Q's being

[1] Storrs McCall, 'Objective Time Flow', *Philosophy of Science*, 43, 1976, pp. 337–62; 'Temporal Flux', *American Philosophical Quarterly*, 3, 1966, pp. 270–81; 'Time and the Physical Modalities', *The Monist*, 53, 1969, pp. 426–46; 'A Dynamic Model of Temporal Becoming', *Analysis*, 44, 1984. R. H. Thomason, 'Indeterminist Time and Truth-Value Gaps', *Theoria*, 36, 1970, pp. 264–81. Vaughan R. McKim and Charles C. Davis, 'Temporal Modalities and the Future', *Notre Dame Journal of Formal Logic*, 17, 1976, pp. 233–38; Robert P. McArthur, 'Factuality and Modality in the Future Tense', *Noûs*, 8, 1974, pp. 283–8; John P. Burgess, 'The Unreal Future', pt 1, *Theoria*, 44, 1978, pp. 157–69.

many-one: if there are many possible worlds accessible from the actual world, it does not follow that there are many possible worlds from which the actual world is accessible.

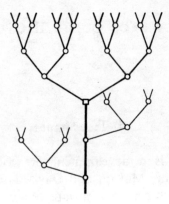

Figure 8:1.

If it be granted that the accessibility relation is in this sense many-one, the structure of accessible worlds will be characteristically tree-like in shape provided we plot time vertically going upwards. Many authors lay the Tree on its side, with time going from left to right;[2] but in Relativity Theory time is always taken as going from bottom to top, and the diagrams look more tree-like if they are vertical.

For purposes of discussion, it is convenient to picture the Tree as discrete, with only a finite number of branches at each junction. Although the mathematics is heavier if we suppose it to be branching continuously with an infinite number of branches, it is not essentially different. Granted this simplification, it is natural to talk of the the linear past of a particular Tree as its 'trunk', all possible future course of events taken together as its 'crown', and the junction at which branching first occurs as its 'node': the node will be indicated by a square box, the other junctions by round ones.

---

[2] As I did, in *A Treatise on Time and Space*, London, 1973, pp. 44–5; and as we have done in our exposition of Reichenbach's analysis of tenses.

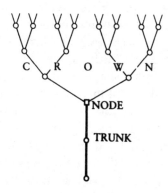

Figure 8:2.

There are many different Trees, all tree-like in structure, but differing in specific shape. These differences correspond on the one hand to different times and on the other to different modalities. The passage of time is pictured as the progressive attenuation of a Tree to some proper sub-tree of itself. Branches are forever falling off, with the trunk getting longer and the node higher, representing the way that with the passing of time possibilities are closed off, but the deposit of assured achievement and indefectible truth increased.[3]

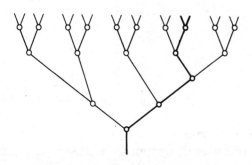

Figure 8:3.

---

[3] See above, ch. 1, p. 9.

The precise shape of a Tree also depends on the modality represented by the accessibility relation Q. For each different sort of possibility, there will be different possible worlds, accessible by virtue of a different relation Q_1, Q_2, Q_3, etc. It may be that the relations are linearly ordered, as logical possibility, physical possibility, biological possibility, human possibility, are: in which case the Trees may be superimposed one on another, as in the figure which shows biological possibility as more restricted than physical possibility. But we should not assume that all modalities can be linearly ordered thus: though in general what is morally possible is more restricted than what is legally possible, it is sometimes morally possible, and indeed necessary, to break the law, and do what is not legally possible.

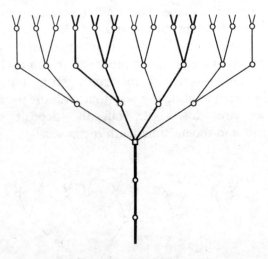

Figure 8:4. Two different modalities give rise to different Trees

Each Tree is a model, which is intended to make clear how temporal truth and temporal necessity can be coherently ascribed to propositions. We are especially concerned with what holds good not just for one Tree, but for all Trees, generalising either over time or over modalities. Theses and rules of inference that hold in all models are said to be 'semantically valid'. We want to show that the logic of temporal standpoints articulated in chapter 6 is semantically valid, that is to say that the logic is 'sound'. There are many other model-theoretic questions that can be raised with regard to Tree Semantics, but these lie beyond the scope of this book.[4]

(ii) Formalities

There are many different ways of defining and of understanding possible worlds. We may take it in an ontologically juicy way, but we do not have to. We may, instead, think of it as a set of contemporaneous[5] atomic propositions or propositional contents. If we have scruples about propositions, we can instead consider well-formed formulae, and take a possible world as some set of well-formed formulae: a set of contemporaneous atomic well-formed formulae or their negations, a maximal consistent set of contemporaneous well-formed formulae, a set of contemporaneous well-formed formulae closed under *Modus Ponens*, or a truth-valuation of contemporaneous well-formed formulae. The phrase 'possible world' should be taken as a term of art. For model-theoretic purposes it does not matter what they are, so long as they can be defined in such a way that they exist and behave in the way required. Intuitively, it is easiest to think of a possible world as characterized by the state of affairs that obtains in it. If we had a questionnaire that went through every propositional content, asking whether it did or did not hold, the answer would tell us what state of affairs obtained at that time; and the set of all the answers would constitute the set of all the possible worlds at that time. We can achieve this mathematically by using a truth-valuation function. A truth-valuation function is

[4] See J. F. A. K. van Benthem, *The Logic of Time*, Synthese Library, 156, Dordrecht, 1983, for an excellent treatment of many technical points.
[5] Here we still make the assumption, which we shall relax in ch. 10, section (vi), that there is no problem in assigning dates to propositions. It would be awkward typographically and opaque as regards exposition to seek complete logical purity at this juncture.

a function onto the two-element Boolean ring, B_2, whose members are TRUE and FALSE. Essentially it *assign* a bivalent truth-value.[6] Each contemporaneous proposition, propositional content, or well-formed formula, is thereby said to *be* either True, or else False. This characterizes a 'possible world'; a different possible world is one in which some proposition *have* a different truth-value.

We have chosen to define possible world in terms of propositions. We could have gone the other way. If we were given contemporaneous possible worlds, we could define propositions as sets of contemporaneous possible worlds. In the figure below we see the Tree of February 28th, 2000, and two intervening stages between the admiral's decision today and what actually happens tomorrow. There are eight possible states of affairs for February 29th, 2000, in only three of which (allowing for a sudden storm's rendering the admiral's decision nugatory) does the sea battle take place. Those three are marked with haloes. And if possible worlds were basic, we could define propositions either as sets, or as unions, of possible worlds.

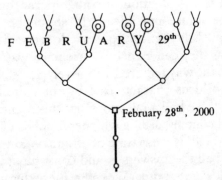

F E B R U A R Y   29th

February 28th, 2000

Figure 8:5.

If we are constructing possible worlds from propositions, we need first to ensure that they are coherent as regards propositional calculus, and then go on to add temporal and modal operators. We start, therefore, with 'valuation rules' for the

[6] See above ch. 4, § (i), p. 60, and § (iv).

sentential connectives of propositional calculus. The valuation function W_i represents the answers to the questionnaire: for each *atomic* proposition, *i.e.* for each propositional content, p_t, $W_i(p_t)$ is defined, and takes either the value TRUE, if the answer to the question is Yes, or the value FALSE, if the answer is No. What rules must we adopt for assigning TRUE or FALSE to complex propositions consistently? Since we can define all the sentential connectives in terms of negation and implication, it is enough to give rules for just these two. Evidently, we should assign false to $-p_t$ when we assign TRUE to p_t, and *vice versa*. The rule for material implication is less intuitively obvious, because material implication does not exactly correspond to the ordinary language use of if . . . then . . ., but if we remember Russell's original definition, $p \rightarrow q =_{df.} -p \vee q$, the rationale of the rule below is clear. A particular valuation function W_i from propositions onto {TRUE, FALSE} characterizes a possible world iff

(1) For any *atomic* proposition, p_t, $W_i(p_t)$ is defined.
(2) For any proposition, p_t,
$W_i(-p_t)$ = FALSE iff $W_i(p_t)$ = TRUE, and
$W_i(-p_t)$ = TRUE iff $W_i(p_t)$ = FALSE.
(3) For any propositions, p_t and q_t,
$W_i(p_t \rightarrow q_t)$ = FALSE iff $W_i(p_t)$ = TRUE and $W_i(q_t)$ = FALSE
$W_i(p_t \rightarrow q_t)$ = TRUE iff $W_i(p_t)$ = FALSE and/or $W_i(q_t)$ = TRUE

The set, w_{it}, of all the propositions at t to which the function W_i *assign* the value TRUE, is a *possible world* (or world, for short) or *state of affairs* at time t; in symbols

$$w_{it} = \{p_t \mid W_i(p_t) = \text{TRUE}\}$$

Though each possible world was initially characterized in terms of just the propositional contents that obtained in it, according to this definition each possible world contains all, and only, the propositions of propositional calculus that follow from those propositional contents.

There are very, very many conceivable possible worlds. We, however, are concerned only with certain subsets of them, those sets which constitute Trees. Trees are relational structures on possible worlds, $\langle W,Q,w_{it}\rangle$, to wit w_{jR} the set of all possible worlds that are either Q- or Q^{-1}-accessible from w_{it}. Formally, the Tree t, of $T(t)$, where w_{it} is the actual world at time t, is defined by the condition

$$T(t) = \{w_{jR} \mid w_{jR}Qw_{it} \vee w_{it}Qw_{jR}\}$$

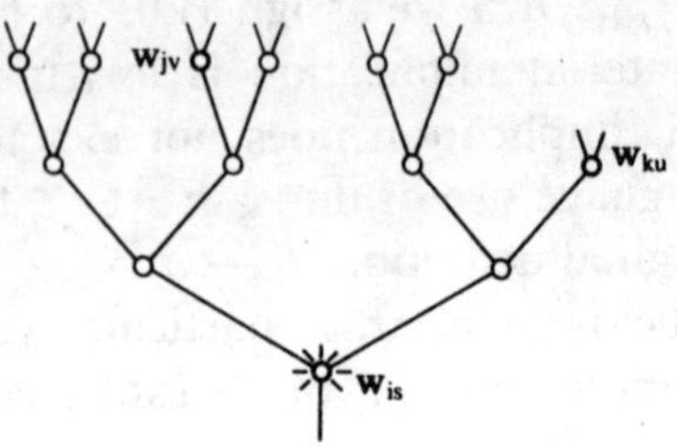

Figure 8:6. For any two worlds, w_{jv} and w_{ku}, there is some world, w_{is}, which is Q^{-1} accessible from both.

Roughly speaking, they are sets of possible worlds linked together by Q-accessibility or Q^{-1}-accessibility, but it is important to note that not every possible world in a particular Tree is Q-accessible or Q^{-1}-accessible from every other possible world, though for any two possible worlds there is a third that is Q^{-1}-accessible from both (Figure 8:6). Every possible world, however, is Q-accessible or Q^{-1}-accessible from every possible world on the trunk, and *vice versa*. We can use this feature, given some Tree of t $T(t)$, to pick out the trunk, and then the node (Figure 8:7): w_{is} is on the trunk of $T(t)$ iff

$$(Aw_{jR})(w_{jR}Qw_{is} \vee w_{is}Qw_{jR});$$

and w_{it} is the node of T iff

$$((Aw_{jR})(w_{jR}Qw_{it} \vee w_{it}Qw_{jR}) \ \&$$

$$(Aw_{ku})(w_{ku}Qw_{it} \ \& \ -w_{it}Qw_{ku} \rightarrow (Vw_{jR})(w_{jR}Qw_{it} \ \& \ -w_{jR}Q \ w_{ku})))$$

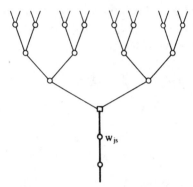

Figure 8:7. But if w_{js} is on the trunk, then every world is either Q-accessible or Q^{-1}-accessible from it.

The second limb of the condition ensures that the node is the Q-most point on the trunk, in that any point that is Q-er than it be not (QvQ^{-1})-accessible from some world in the Tree.[7]

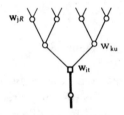

Figure 8:8. w_{it} is the node iff (i) it is on the trunk, and (ii) any world, w_{ku} other than itself which is Q-accessible from it, is not on the trunk.

Other definitions are possible. We can define first branches, as maximal Q-chains of possible worlds, and then a Tree as the union of all those branches that go through its node. Intuitively, a branch is a possible world-history, and a Tree is all the

---

[7] It should be noted that though w_{jR} occurs in both limbs of the definition, it is bound by separate quantifiers. Some readers may prefer to replace it on one occasion of use by w_{lv} with the proviso that there is no restriction on the temporal order of v with respect to the other indices.

world-histories that are still possible. In spite of the simplicity of this approach, there are reasons for not adopting it, which will emerge in the next chapter.[8]

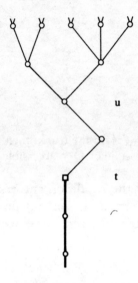

Figure 8:9.

Having defined Trees, we can define sub-trees in various ways. For any given Q, a sub-tree of a Tree of t $T(t)$ is a tree-like structure whose set of possible worlds is a sub-set of the set of worlds that are members of T, and a proper sub-tree of T is a tree-like structure whose set of possible worlds is a proper sub-set of the set of worlds that are members of T. Alternatively, a sub-tree(u) of the Tree of t $T(t)$ is a Tree whose node w_{ku} is Q-accessible from w_{it}, the node of $T(t)$;[9] and a proper sub-tree(u) of the Tree of t $T(t)$ is a Tree whose node w_{ku} is Q-accessible, but not Q^{-1}-accessible, from w_{it}, the node of $T(t)$ (Figure 8:9). Other definitions are possible in terms of the trunk

---

[8] Ch. 9, § (iii), pp. 168–9.
[9] It is here that we need to have Q reflexive, in order that the formalism should not become awkwardly cumbersome; see above, ch. 5, § (iv), p. 94.

of one including the trunk of the other, or, when we have
defined a branch, in terms of all the branches of the one being
branches of the other. Just as a set is the union of all its sub-sets,
so a Tree is the union of all its sub-trees: any member of a
sub-tree(*u*) of the Tree(*t*) is a member of Tree(*t*), and any
member of Tree(*t*) is a member of some sub-tree(*u*).

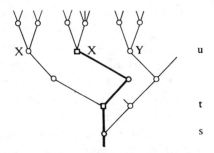

Figure 8:10.

Besides the actual Tree(*u*) there are many other *possible*
sub-trees whose nodes are states of affairs at time *u*; e.g. those
marked X and Y. These are in some sense possible sub-trees at
u, but we need to be careful in specifying the date of the
possibility. There is no longer any possibility of them at time *u*,
as things have turned out, although there was a possibility of X
at date *t*, and of both X and Y at date *s* (Figure 8:10). That is to
say, there was a possibility of X_u at *t*, and a possibility of X_u and
of Y_u at *s*.

(iii) Tree Satisfaction and Temporal Truth

The Tree of *t* T(*t*) gives us a model of temporal truth. Granted
some particular Tree of *t* T(*t*), we can ask whether it satisfies
any particular proposition, p_R. We say that it does if p_R *be* True
in every possible world, w_{jR}, of date r. Symbolically, we use the
double turnstile, and say

$$T(t) \vDash p_R.$$

If w_{it} be the actual world at time t, and Tree $\mathbf{T}(t)$ its Tree, we say that p_R is (temporally) true at time t, or $\text{True}_t(p_R)$,[10] iff, in every world w_{iR} that is Q-accessible or Q^{-1}-accessible from the node of the Tree of t, p_R be (timelessly) True. And similarly, we say that p_R is false at time t, or $\text{Fls}_t(p_R)$, iff in every world w_{iR} that is Q-accessible or Q^{-1}-accessible from the node w_{it} of the Tree of t $\mathbf{T}(t)$ p_R be False.

If $s \leq t$, there is only one world w_{js} that is Q-accessible or Q^{-1}-accessible from the node t. In this world according to the valuation function W_j either p_s *be assigned* the truth-value TRUE, in which case p_s is temporally true at t, $\text{True}_t(p_s)$, or p_s *be assigned* the truth-value FALSE, in which case p_s is temporally false at t, $\text{Fls}_t(p_s)$. It follows that for temporal truth about the present and past the Principle of Bivalence holds. If $t < u$, there is more than one world that is Q-accessible from the node of t: if it so happens that p_u *be* True in all of them, then p_u is temporally true at time t. Equally, if it so happens that p_u *be* False in all of them, then p_u is temporally false at time t. But it may be that p_u *be* True in some worlds that are Q-accessible from the node of the Tree and False in others, in which case it is neither true nor false at time t. That is, the Principle of Bivalence does not always hold for temporal truth about the future.

Temporal truth has been represented in terms of Trees, and Trees are related by the sub-tree relation. Any proposition, p_R, which is temporally true at time t, *be* True in every world w_{jR} that is a member of the Tree of t $\mathbf{T}(t)$: but any member of any sub-tree of the Tree $\mathbf{T}(t)$ is a member of the Tree $\mathbf{T}(t)$; so if p_R *be* True in every member of the Tree $\mathbf{T}(t)$, it *be* True in every member of the sub-tree(u) of the Tree $\mathbf{T}(t)$. Hence indefectibility. What is temporally true at time t must remain true for all subsequent time, because in every subsequent sub-tree the set of possible worlds is not being increased, and so there is never any further possibility of what is temporally true becoming false. Similarly, there is never any further possibility of what is

---

[10] Although '$\text{True}_t(p_R)$' has an initial capital, it ascribes temporal truth, notwithstanding the convention of ch. 4, § (i), p. 58.

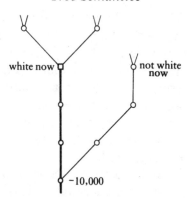

Figure 8:11. Showing Tree$_{-10000}$ and Tree$_{now}$. What is True$_{now}$ need not have been True$_{-10000}$.

temporally false becoming true, so that temporal falsity is likewise indefectible. Indefectibility is not omnitemporality. Propositions may become true or false, but this does not imply that they were all along true or false. If something is white now, it is$_{now}$ indefectibly true that it is white now, since in every now-possible now-world, namely the one and only actual now-world, it is white: τὸ μὲν οὖν εἶναι τὸ ὅν ὅταν ἦ, ἀνάγκη [*to men oun einai to on, hotan ei, ananke*] (that which is, necessarily is) when it is.[11] But it was not temporally true ten thousand years ago that it would be white, because then there were many Q-accessible worlds in which it would not have been white now. What is true now always will be: but it does not follow that it always was.

We can in a similar way meet an argument of Kenny's that 'whatever changes of plan we may make, the future is what takes place after all the changes are made; what we alter is not the future, but our plans; the real future can no more be altered than the past'.[12] Kenny is adopting a future temporal

[11] See above, ch. 7, § (i), p. 124, n.5.
[12] Anthony Kenny, *Aquinas: A Collection of Critical Essays*, London, 1969, p. 267. Kenny has developed his position since, and may no longer be open to the criticism expressed here.

standpoint. From the standpoint of tomorrow, u, either it will be true then that there is a sea battle or it will be false then that there is a sea battle: $True_u p_u \vee Fls_u p_u$; it is then unalterable. But if it is true then, it does not follow that it is true now; $True_u p_u$ does not imply $True_t p_u$: equally, if it is false then, it does not follow that it is false now; $Fls_u p_u$ does not imply $Fls_t p_u$. It is not unalterable now. Kenny's argument depends on an illegitimate backwards shift of the temporal standpoint from which temporal truth or falsity are being assessed.

Temporal truth can be iterated. We need to be careful about iterating any sort of truth on account of Tarskian difficulties,[13] but the concept of temporal truth is sufficiently limited to pose no problems. Certainly I can intelligibly ask whether it is true now that it was true ten thousand years ago that it is white now. According to Tree Semantics, we should interpret this as a question now about the world of ten thousand years ago, whether in that world $True_{-10000}(white_{now})$ held good. We can construe the informal account we have given thus far as a formal valuation rule for the unary predicate $True_t$:

(4) $W_i(True_t(p_R)) = \text{TRUE}$ iff $T(t) \vDash p_R$ and
  $W_i(True_t(p_R)) = \text{FALSE}$ iff $T(t) \vDash -p_R$, and
  $W_i(True_t(p_R))$ be not defined otherwise.

In that case, the question of truth ten thousand years ago becomes a question of the Tree of ten thousand years ago, and how it stands in relation to the Tree of now. When the date of the iterated temporal truth is the same as, or after, the temporal truth being considered, it is a fairly vacuous operator:[14] it is true now that it was true ten thousand years ago that it is white now iff it was true ten thousand years ago that it is white now. But it is not so vacuous when the iterated temporal truth is before the one being considered: it is not the case that it was true ten thousand years ago that it is true now that it is white iff it is true now that it is white. Kenny's argument, too, seems harder to

---

[13] See above, ch. 4, § (i), p. 59.
[14] See above, ch. 4, § (i), p. 60.

resist when posed in terms of iterated temporal truth. Today it is true that tomorrow it will be either true that there is a sea battle or else false: and so it seems that today too it is either true or else false that there will be a sea battle. Kenny's illegitimate backwards shift of the temporal standpoint of temporal truth or falsity is made more tempting because it is nested within an earlier temporal truth. Because in $\text{True}_t(\text{True}_u(p_u) \vee \text{Fls}_u(p_u))$ the time of unalterability, contemporaneous with the event p_u, is nested within temporal truth assessed from the standpoint t, it is easy to suppose that it implies an unalterability contemporaneous not with the event but the temporal standpoint, t, of the temporal truth. Aristotle and Kenny may have been susceptible to covert pressure from iterated temporal truth, sometimes vacuous and sometimes viciously non-vacuous.

(iv) The Logic of Temporal Standpoints

We can give a very similar account of the omnitemporal[15] truth of RE propositions, and hence of the validity of the logic of temporal standpoints put forward in chapter 6. Granted some Tree $T(t)$ with node at time t, an RE proposition of the form $[p_R]_t$ is True iff p_R *be* True in every w_{iR} that is in the Tree $T(t)$; that is to say, in every w_{iR} that is Q-accessible or Q^{-1}-accessible from the node of the Tree $T(t)$: and an RE proposition of the form $[p_R]_t$ is False iff p_R *be* False in the Tree $T(t)$; that is to say, in every w_{ir} that is Q-accessible or Q^{-1}-accessible from the node of the Tree $T(t)$. It is evident that if $R \leq t$, the Principle of Bivalence holds, and every proposition of the form $[p_R]_t$ *be* either omnitemporally True or else omnitemporally False: but the Principle of Bivalence does not always hold when $t < R$, because since there may be more than one possible world in the Tree $T(t)$, p_R may *be* True in some of them and False in others, so that $[p_R]_t$ *be* neither True nor False.

Thus far we have dealt with only one Tree $T(t)$, and shown how it would assign truth and falsity to RE propositions. But

---

[15] See above, ch. 4, § (i), pp. 57–8.

little depended on the particular choice of Tree, and we can generalise to consider whether the rules of inference and axioms of the logic of temporal standpoints hold for all Trees.[16]

We consider first the inference from $[p_R]_t$ to $[p_R]_u$ where $t \leq u$. The condition under which $[p_R]_t$ be True in any particular Tree $T(t)$ is that p_R be True in every world w_{jr} that is in Tree $T(t)$, in which case it must *be* True in every world w_{jR} that is in a sub-tree(u), and that is the condition for $[p_R]_u$ to *be* True. The inference from $[p_R]_t$ to $[p_R]_u$ thus holds in any particular Tree, and hence in every Tree. We thus have validated

R2 $[p_R]_t \vdash [p_R]_u$ provided $t \leq u$.

Axiom 1 $\vdash p_t \leftrightarrow [p_t]_t$

is clearly valid, for in any Tree of t $T(t)$, if p_t *be* True, then since there is only one possible world at the node and p_t *be* True in that one, it *be* True in all Q-accessible or Q^{-1}-accessible worlds of date t, which is the condition for $[p_t]_t$ to *be* True; and conversely, if $[p_t]_t$ *be* True, then p_t *be* True in all Q-accessible or Q^{-1}-accessible worlds of date t, and hence in the node, so that p_t *be* True. Hence

$$p_t \leftrightarrow [p_t]_t$$

is true in every Tree of t, and so

$$p_t \leftrightarrow [p_t]_t$$

is valid, or in symbols,

$$\vDash p_t \leftrightarrow [p_t]_t.$$

We argue in a similar vein that the axiom κ is valid.

κ $\vdash [p_R \to q_r]_t \to ([p_R]_t \to [q_R]_t)$

holds in each and every Tree $T(t)$ with node at time t. For in any such Tree $T(t)$, if $[p_R \to q_R]_t$ *be* False or if $[p_R]_t$ *be* False, then

---

[16] See above, ch. 6, § (i), pp. 109–10.

$$[p_R \to q_R]_t \to ([p_R]_t \to [q_R]_t)$$

be trivially True: while, on the other hand, the condition for $[p_R \to q_R]_t$ and $[p_R]_t$ *being* True, is that $p_R \to q_R$ and p_R both *be* True in every world w_{jR} of date R in that Tree $T(t)$, in which case in each one of them it follows by *Modus Ponens* that q_R *be* True; and if q_R *be* True in every world w_{jR} of date R in the Tree of t $T(t)$, then $[q_R]_t$ *be* True. So, either way,

$$[p_R \to q_R]_t \to ([p_R]_t \to [q_R]_t)$$

be True in every Tree of t $T(t)$, and hence is valid, or in symbols:

$$\vDash [p_R \to q_R]_t \to ([p_R]_t \to [q_R]_t)$$

It is not strictly necessary to check R3',[17] since it turned out to be derivable from the other rules and axioms; but it is none the less instructive to see how it works out. It states

R3' $[p_t]_v \vdash [p_t]_u$ even though $u < v$, provided $t < u$.

Granted that $t \le u$, p_t *be* True in the Tree of u $T(u)$ with node at u, iff p_t be in w_{it} where w_{it} be a possible world on the trunk of Tree of u $T(u)$. But if w_{it} *be* on the trunk of Tree of u $T(u)$, it must also *be* on the trunk of the Tree of v $T(v)$. Although the Tree of u is not a sub-tree of the Tree of v, the trunk of u is a sub-trunk of the trunk of v. So, once again, since this holds of Trees generally, R3' is valid.

The logic of RE propositions is very similar to that of temporal truth. Each is made up of three constituents: dated propositional contents, a second set of dates, and the two truth-values, TRUE and FALSE. For the concept of temporal truth the second set of dates is associated more closely with the truth-values, to yield a concept of temporal truth at a particular time: for the RE calculus the second set of dates is associated more closely with the propositional contents, to yield Reichenbachian propositions. The Rule of Inference **R2**

[17] See above, ch. 6, § (i), p. 111.

corresponds to the indefectibility of temporal truth, and axiom 1 corresponds to the vacuity and therefore iterability of contemporaneous temporal standpoints.

(v) Temporal Modalities

We can use Trees to construct valuation rules for temporal necessity and possibility, which we shall symbolize by L_t and M_t.[18] The valuation rules are essentially similar to those for temporal truth, except that where temporal truth was assigned no truth-value, temporal necessity is assigned FALSE and temporal possibility is assigned TRUE.

We *assign* TRUE to $L_t p_R$, iff **every** world w_{ir} of date R in the Tree $T(t)$ with node at t is one in which p_R *be* True,
and otherwise we assign FALSE to $L_t p_R$.
Similarly we assign TRUE to $M_t p_R$, iff **some** world of date R in the Tree $T(t)$ with node at t is one in which p_R *be* True,
and otherwise we assign FALSE to $M_t p_R$.
Granted this valuation for L_t, we consider the Rule of Necessitation,[19] here expressed as

RL
$$\text{If } \vdash p_R \text{ then } \vdash L_t p_R.$$

If p_R is a theorem, then it holds in every possible world w_{ir} of date R; it therefore holds in every possible world w_{ir} of date R in the Tree of t $T(t)$. But this is the condition for $L_t p_R$ being True. Therefore, $L_t p_R$ *be* True in every Tree of t $T(t)$, and hence $\vdash L_t p_R$. So the rule of inference RL holds for all Trees T, and hence is valid.

We argue next that Axiom K, characteristic of all modal systems, is valid:

K
$$\vdash L_t(p_R \rightarrow q_R) \rightarrow (L_t p_R \rightarrow L_t q_R).$$

---

[18] Note again that L is no longer being reserved for logical necessity, as it was in ch. 3, § (iv), p. 41.
[19] See above, ch. 5, § (i), pp. 82–3.

The argument is exactly similar to that with the temporal standpoint operator [ ]$_t$ in section (iv) above. If $L_t(p_R \rightarrow q_R)$ or $L_t p_R$ be False in any Tree $T(t)$ with node at time t, the thesis holds trivially in that Tree. If $L_t(p_R \rightarrow q_R)$ and $L_t p_R$ be True, then in that Tree $p_R \rightarrow q_R$ and p_R be True in every world w_{jR} with date R, which is the condition for $L_t q_R$ *being* True in that Tree. So the thesis K *be* True in every Tree, and is therefore valid.

Temporal necessity is closely related to temporal truth: it is what temporal truth becomes if we insist on the Principle of Bivalence, and assign a truth-value, TRUE or FALSE, to every proposition.[20] Temporal truth closely parallels the logic of temporal standpoints, expressed in the calculus of RE propositions. We may ask what is the relation between temporal necessity and RE propositions? As we have defined them, RE propositions do not always obey the Principle of Bivalence: when $t < u$, $[p_u]_t$ may be assigned no truth-value. But we are under pressure from the Principle of Bivalence, and may feel obliged to say that if it *be* not True that there is going to be a sea battle tomorrow, for the admiral has not yet made up his mind, it must *be* False. In that case we may be importing the sense of necessity not into temporal truth, but into the tense operator [ ]$_t$, and turning 'is going to' into 'is to' or 'surely shall'.[21] Such pressure should be resisted. For one thing, we want to keep the modal logic of the future operator – it does not commute with negation, and does not divide under disjunction – distinct from necessitarian overtones: for another, there are important differences between temporal necessity and other tense operators.

The valuation rules for of temporal necessity and possibility were given in terms of possible worlds in a Tree $T(t)$, but could have equally well been expressed in terms of accessibility. We could have said:

We assign TRUE to $L_t p_R$, iff every Q-accessible and Q^{-1}-accessible world w_{iR} of date R in the Tree $T(t)$ with node at t is one in which p_R be True,

---

[20] See above, ch. 4, § (iv), pp. 74, 77–8.
[21] See above, ch. 4, § (iv), p. 78.

and otherwise we assign FALSE to $L_t p_R$.

Similarly we assign TRUE to $M_t p_R$, iff some Q-accessible or Q^{-1}-accessible world of date R in the Tree $T(t)$ with node at t is one in which p_R *be* True,

and otherwise we assign FALSE to $M_t p_R$.

The words 'Q-accessible or Q^{-1}-accessible world', though redundant, make it clear that the accessibility relation of temporal necessity and temporal possibility is neither Q nor Q^{-1}, but $(Q \vee Q^{-1})$ which is not transitive. In this important respect L_t differs from G and H.

Since $(Q \vee Q^{-1})$ is symmetrical, L_t satisfies the Browerian axiom, $\vdash p \to LMp$, or equivalently, what is possibly necessary is TRUE. Since temporal necessity is not S4-like, it is not S5-like. The thesis $M_t L_u p_R \to L_t p_R$ does not hold. A counter-example is shown in the diagram. $L_u p_R$ holds on the right-hand branch but not the left-hand one, and so $M_t L_u p_R$ be True in this Tree of t $T(t)$ without $L_t p_R$ being True.

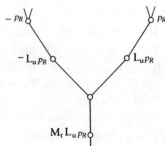

Figure 8:12.

(vi) Moods and Tenses

Tree Semantics explains why in many languages a past tense is often used with modal effect, and in particular in conditional sentences to indicate possibility rather than actuality.

In English the subjunctive mood has all but disappeared, its place being taken by auxiliary verbs in the past tense – *should, would, could, might, ought.* The effect of the past tense is to make the modal force less personal or debatable. 'I ought' is 'I owed'. Being a past commitment, it is now unalterable. At one time, perhaps, I could have got out of it, but now that I have taken the loan, entered into the contract, accepted the engagement, I am irretrievably bound. Similarly, 'I should' is more obligatory than 'I shall'; 'I shall' rests only on my own decision, whereas 'I should' is something given, and as of now unalterable. It is not just me deciding, but something outside my control, an obligation, which I could choose to ignore, perhaps, if I were that way inclined, but which it is not up to me to alter or abridge.

In many languages the past tense is used to indicate some sort of unactual possibility. In Latin the subjunctive mood is also used, and in Greek the subjunctive and optative moods, but the optative has a marked flavour of a past tense as well as being more remote from actuality. In French the conditional is obviously past in flavour, being formed from the imperfect of *avoir*. In Spanish and Portuguese a conditional is called *futuro do pretereto*, a 'future of the past'. In English we use *should* and *would* in counter-factual conditionals, and in the antecedent clauses the apparent past-tense forms *If I were to, If I were, If I was, If I had, If I had had*. These are not genuine past-tense forms, for we can use them with the present indexicals *now* and *today* – *If I were in Rome today, I should see the Forum, si essem Romae hodie, viderem Forum.* Tree Semantics enables us to see both how a past tense can do duty for a non-indicative mood, and the limitations of that approach, which led Latin and Greek to make use also of the subjunctive and optative moods.

If we consider the Latin conditional forms, it is evident that the further back we have to go, to find a state of affairs from which could have evolved the possibility envisaged in the antecendent, the more remote that possibility is from actuality. *Si uxorem habeo, eam amabo* (Figure 8:13) is an Open Conditional looking forward to a future possibility without any indication as to whether or not it will actually occur. *Si uxorem habeam, eam amem* (Figure 8:14) is a Subjunctive Hypothetical – *If I were to have a wife* is a very small step back immediately cancelled by a looking forward, and points towards my not having a wife, while leaving open the possibility of my having one. *Si uxorem haberem*, the rising executive muses, *eam amarem* (Figure 8:15). It is a Contrary-to-fact conditional. He has not got a wife; if he had, he would not say *If I had*, but, significantly changing the tense, *Since I have a wife* . . . . *If I had* is a step back, not as small as *If I were to*, and indicates the definite unfactuality of his being married now, not just a present tendency towards celibacy. But though wifelessness is his present state, it need not be forever, whereas the pluperfect, going back many steps, expresses an Impossible Conditional. *Si uxorem habuissem*, the old bachelor says, *eam amavissem* (Figure 8:16) – once, long ago, there was a girl I might have

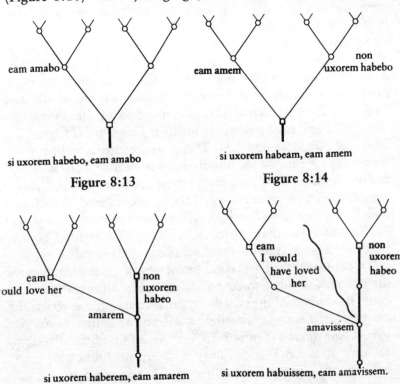

eam amabo

si uxorem habebo, eam amabo

Figure 8:13

eam amem — non uxorem habebo

si uxorem habeam, eam amem

Figure 8:14

eam ould love her — non uxorem habeo

amarem

si uxorem haberem, eam amarem

Figure 8:15.

eam I would have loved her — non uxorem habeo

amavissem

si uxorem habuissem, eam amavissem.

Figure 8:16

married, but that possibility is irretrievably closed. [The *amavissem* is also dated to the past, but the English takes us forward to the present and back again: she would now be in the state of having been loved by me.]

We should compare the English use of tenses to give a measure of the remoteness of a possibility envisaged, in terms of how far back one would have to go to find that possibility still open with the account of counterfactuals given by David Lewis,[22] in which he supposes a general measure of remoteness of possible worlds as compared with the actual one. The temporal measure is much more restricted than his. However far back one went in time, the laws of nature would still hold good, and there never was a time when pigs could fly or perpetual motion machines could work. We can suppose laws of nature different from what they are, together with many other possibilities which never were at any time temporal possibilities. Our use of tense to express unactuality is illuminating, but does not exhaust the ways in which possibilities can be remote.

English usage is, in some cases, unhappy. In expressing a future tense counterfactual *If I were to have a wife* by a past tense it is having to indicate unactuality by supposing the future to be present in its causes and back-tracking from the present tendency. But some futures are not present in their causes. The distinction between those futures that are possible but will not be actual, and the future that will be actual, may be one that will only emerge in the event. A subjunctive rather than a suppositious past is called for, in order adequately to distinguish future actuality from future possibility. The difficulty Tree Semantics has in accommodating that distinction indicates the limitations of our present approach.

[22] David Lewis, *Counterfactuals*, Oxford, 1973; and 'Counterfactual Dependence and Time's Arrow', *Noûs*, 13, 1979, pp. 455–76. Robert Stalnaker, 'A Theory of Conditionals', in Ernest Sosa, ed., *Causation and Conditionals*, Oxford, 1975, pp. 165–79. Storrs McCall, 'Counterfacutals Based on Real Possible Worlds', *Noûs*, 18, 1984, pp. 464–77. For criticisms, see Jonathan Bennett, *Canadian Journal of Philosophy*, 4, 1974, pp. 381–402.

(vii) Critique

Tree Semantics has many merits. It gives an intuitively persuasive picture of time and the passage of time. It correlates different theses in tense logic with different topologies of time or the course of events. It gives an account of temporal truth, the truth of RE propositions, and temporal necessity, which brings out the connexions between them.

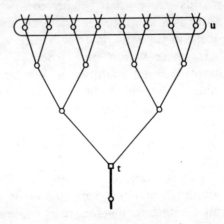

Figure 8:17. If $T(t)$ has all its worlds at u containing p_u, then the subtree, marked heavier, does so too.

Temporal truth is indefectible, and not subject to the Principle of Bivalence. Tree Semantics shows the reasons. If all the members of a set do *have* a certain feature, then all the members of any sub-set *must have* it too, although if not all the members of a set *have* a certain feature, all the members of a sub-set *may have* it none the less. So if the Tree of t $T(t)$ *have* all its possible worlds at date u having a certain feature (8:17), *e.g.* that a sea battle *be* taking place, then all its sub-trees must have that feature too: but if it *do not have* all its possible worlds at date u *having* a certain feature, one of its sub-trees can nonetheless *have* all its possible worlds *having* the feature of a sea battle taking place. Thus temporal truth may be acquired,

but once acquired, cannot thereafter be lost. Hence indefectibility. Moreover, if the Tree of *t* *do not have* all its possible
worlds at date *u* *having* the feature that a sea battle *be* taking
place, it does not follow that none of its possible worlds at date
u *have* the feature that a sea battle *be* taking place. If, as is
natural, we want p_u to be temporally FALSE at t iff $-p_u$ is
temporally TRUE at *t*, it follows that p_u will be neither temporally
TRUE at *t* nor temporally FALSE at *t* when the Tree of *t* do not
have either all or none of its possible worlds at date *u* *having* the feature that a sea battle *be* taking place. Hence truth-
gaps in the ascription of temporal truth.

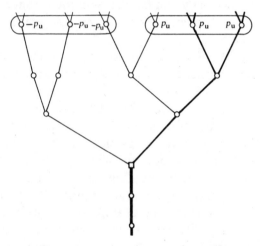

Figure 8:18. But if **T**(*t*) does not have its worlds at *u* containing p_u,
a subtree (the one on the right) may none the less have all *its*
worlds at *u* containing p_u.

Tree Semantics shows the close parallel between temporal
truth and temporal necessity, as also their significant differences. It is correct to say that a proposition is temporally
necessary at a given time only under the same conditions as its
being correct to say that it is temporally TRUE at that time, and
whenever a proposition is temporally FALSE at *t* it is correct to
say that it is not necessary; but it is also correct to say that it is
not necessary when it is neither TRUE not FALSE. What is a

truth-gap where temporal truth is concerned becomes a further case of falsity in assessments of temporal necessity. Temporal necessity lacks the negation-symmetry of temporal truth and falsity. Whereas p_u is temporally false at t only if $-p_u$ is true, p_u may be not temporally necessary at t though $-p_u$ is not temporally necessary either. Tree Semantics shows the reason to be that temporal truth and falsity are essentially contraries, being defined in terms of all and none of the accessible worlds of the relevant date, whereas necessity and non-necessity are contraries, being defined in terms of all and not all. And so, whereas temporal falsity is, like temporal truth, indefectible, temporal non-necessity can be lost with the passage with time; or to put it terms of possibility, temporal possibilities can be lost but never acquired.

In spite of its merits, the account we have given of Tree Semantics has defects. It gives no account of the simple future tense and valedictory truth. It has been developed with respect to only one modality, and the defeasible ascription of indefectible truth to predictions essentially involves more than one modality. Moreover, it still gives only a rather static picture of time. Although it allows for the arbitrary choice of temporal standpoint represented by Reichenbach's reference point R, it does not adequately capture the mandatory change of the present moment of speaking, represented by Reichenbach's S. We have spoken of *the* Tree $T(t)$ with node at t, but from the standpoint of earlier times, say s, there are many sub-trees of s with nodes at t, and we have indicated no way of distinguishing the actual Tree $T(t)$ with node at t from merely possible sub-trees of some earlier Tree. Tree Semantics is in danger of being either too static, or begging the question of temporal change – if it fails to record the fact that the world Tree is constantly shedding its branches, it gives only a frozen picture, and if it is to accommodate the change of modal status in its sub-trees, more needs to be said about time and modality.

9

The Actual Future

(i) Actuality

Although Tree Semantics goes a long way to filling out an exact and detailed account of our different uses of moods and tenses, and the logic of temporal truth, there remains a sense, which has been articulated by some recent writers, that the account so far given is not adequate.[1] For one thing, we have not as yet given an adequate account of the truth of future-tensed utterances, whether conjectures aspiring to valedictory truth or predictions aspiring to something more substantial. In the second place, there was some unhappiness in representing shades of unactuality by means of tenses alone.[2] More generally, questions about actuality were begged in the last chapter. Trees were assumed to be actual, but not sufficiently distinguished from sub-trees which represented mere possibilities: the Tree of t was a sub-tree of the Tree of s, but it was not clear which sub-tree was the actual one as opposed to many others that had been possible at the time s but had failed to come to pass. The essential difference between the actual Tree of t and merely possible sub-trees of the Tree of s is that the node of the actual Tree is the actual present represented by

[1] Robert P. McArthur, 'Factuality and Modality in the Future Tense', Noûs, 8, 1974, pp. 283–8. His argument is countered by Vaughan R. McKim and Charles C. Davis, 'Temporal Modalities and the Future', Notre Dame Journal of Formal Logic, 17, 1976, pp. 233–8. See further, John P. Burgess, 'The Unreal Future', pt 1, Theoria, 44, 1978, pp. 157–69.
[2] See above, ch. 8, end of § (vi), p. 157.

Reichenbach's S, whereas the nodes of sub-trees are temporal standpoints we can, if we wish, take up, but do not have to, represented by Reichenbach's R. We cannot choose what time it is now: it changes whether we will it or not, and the actual Tree changes correspondingly, shedding branches as time goes on, with the trunk ever longer and the node ever higher. Tree Semantics needs to consider not just a single Tree, but the relation between a succession of Trees, each a sub-tree of its predecessors.

(ii) White Now

The difficulty becomes acute when we seek to provide some semantics for the for the simple future (S–R,E) tense. Neither of the accounts offered in the last chapter will serve. If we say that a future proposition is true if it *be* True in every Q-accessible world of the relevant date, we are assimilating the simple future to the posterior present, and futurity to temporal necessity. If, on the other hand, we say that a future proposition is true if it *be* True in some Q-accessible world of the relevant date, we seem to be making the future *too* tenuous and insubstantial, and failing to represent the actuality of the actual future, something more than possibility, a bare may-be, without being a necessity, an inevitable must-be. Future trees are not actual Trees, and to give an account of future truth in terms of future trees is not actually to give a full account of future truth. We can see this most clearly if we consider again Aristotle's 'white now' argument, *if something is white now* – which we symbolize by p_u[3] –, *it was true (in some sense of 'true') to say beforehand that it will be white.*[4] Although we want to avoid an illegitimate slide from the simple future into the posterior present, it is hard to deny that if p_u is the case now then it was the case that it will be

---

[3] In the 'white now' argument, the subscript u represents now (yesterday's tomorrow), rather than tomorrow.

[4] See above, ch. 7 § (i), p. 120.

the case that p_u, just as if p_u is the case now then it will be the case that it was the case that p_u. These two 'mixing axioms' have great appeal,[5] but the former cannot be rendered in terms of the RE calculus, or brought out as valid in Tree Semantics as thus far developed.

Let us use F_t for the 'actual future', expressed by means of the simple future (S–R,E), and instead of the posterior present (S,R–E) 'there is going to be . . .', let us use the modal analogue 'there has got to be . . .', which we symbolize as $[F]_t$. $[F]_t p_u$ is a special case of $L_t p_u$, when $t < u$, and p_u is some proposition, not necessarily atomic. It has the merit of always having a truth-value TRUE or FALSE,[6] and suggests the further symbol $\langle\!\!\!F\rangle_t$ 'there may be . . .', defined, analogously to M_t, as $-\boxed{F}_t-$.[7] Let us also use P_t for the past; no distinction needs to be made between the actual and the necessary or possible past.

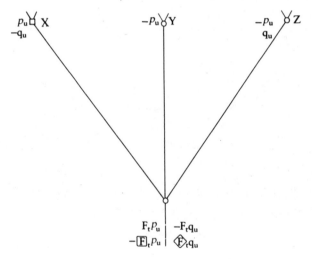

Figure 9:1.

[5] See above ch. 5, § (iv), p. 96.

[6] See above, ch. 4, § (iv), pp. 77–8, and ch. 5, § (ii), p. 91.

[7] This terminology is adapted from McKim and Davis, 'Temporal Modalities' pp. 233–8, but I have added subscripts throughout, for the reasons given in ch. 5, § (v), pp. 98–9.

McArthur gives a convincing argument for distinguishing **F** from both $\boxed{F}$ and $\Diamonddot{F}$. If p_u *be* True in world X, but not in Y or Z, then $F_t p_u$ *be* True, while $\boxed{F}_t p_u$ *be* False, and if q_u *be* False in X, but not in Z, then $\Diamonddot{F}_t q_u$ *be* True, while $F_t q_u$ *be* False.[8] It is easy to state truth conditions for $\boxed{F}_t p_u$ formally in terms of Trees. $\boxed{F}_t p_u$ *be* True iff $t < u$, and every valuation function on the Tree of t assign TRUE to p_u; and otherwise $\boxed{F}_t p$ *be* False. Similarly $\Diamonddot{F}_t p_u$ *be* True iff $t < u$, and some valuation function in the Tree of t *assign* TRUE to p_u, and otherwise $\Diamonddot{F}_t p_u$ *be* False. Evidently $\boxed{F}_t$ is not a modally vacuous operator: for it is not a theorem that.

$$-\boxed{F}_t p_u \to \boxed{F}_t - p_u$$

nor that

$$\boxed{F}_t (p_u \vee q_u) \to \boxed{F}_t p_u \vee \boxed{F}_t q_u.$$

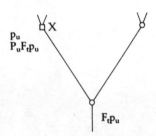

Figure 9:2.

in particular, as we have seen, although $\boxed{F}_t(p_u \vee -p_u)$ is a theorem, $\boxed{F}_t p_u \vee \boxed{F}_t - p_u$ is not. With regard to the analogues of Prior's mixing axioms we have as a theorem

$$\vdash P_u \boxed{F}_t p_u \to p_u$$

but not the converse

$$p_u \to P_u \boxed{F}_t p_u.$$

With $\Diamonddot{F}_t$ the position is reversed: we have as a theorem

---

[8] McArthur, 'Factuality and Modality' pp. 285–7. See Figure 9:1.

$$\vdash p_u \to P_u \langle\!\!\text{\small F}\!\!\rangle_t p_u$$

but not the converse

$$P_u \langle\!\!\text{\small F}\!\!\rangle_t p_u \to p_u.$$

With the actual future operator, F_t, we want the two theorems of modal vacuity to hold:[9]

Commutativity with Negation $\vdash -F_t p_u \to F_t - p_u$

Division under Disjunction $\vdash F_t(p_u \lor q_u) \to (F_t p_u \lor F_t q_u),$

and we want the mixing axiom to hold for the 'white now' case:

$$\vdash p_u \to P_u F_t p_u, \text{ and conversely, } \vdash P_u F_t p_u \to p_u.$$

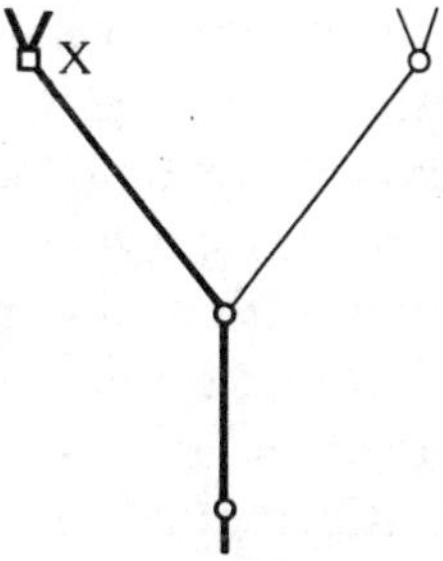

Figure 9:3. 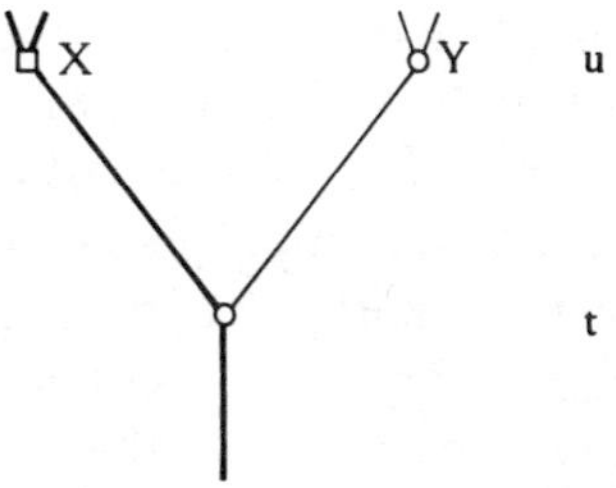   Figure 9:4.

Let us consider the 'white now' argument first. If it is white now, at time u, then the X branch is one in which every valuation *assign* the truth-value TRUE to its being white at u. In this branch it would be the case that every valuation at time t was one which also *assign* TRUE to p_u. So, *relative* to this branch, $F_t p_u$ *be* True at t; i.e. $F_t p_u$ was, so far as this branch *go*, temporally true at t. So, relative to this branch, $P_u F_t p_u$ *be* True at t ; i.e. $P_u F_t p_u$ is, so far as this branch *go*, temporally true at u. But this *is the actual* branch. Although there were other possible branches of the Tree of t, each with nodes at u – e.g. that marked Y in the diagram – they are no longer relevant. The actual Tree is the one that has grown along the X branch. Hence

---

[9] See above, ch. 5, § (i), p. 89.

$P_uF_tp_u$ is a semantic consequence of p_u in the relevant model. Conversely, the only way for $P_uF_tp_u$ to *be* True in the X branch is for p_u to *be* True also. Hence p_u is equally much a semantic consequence of $P_uF_tp_u$ in the relevant model. And this is generally true: the proposition

$$p_u \rightarrow P_uF_tp_u$$

holds good in each and every relevant model, and so is semantically valid.

'But', we protest, 'is it fair to exclude the other branches, representing other possible world histories?' Although, now, *ex post facto*, we dismiss them as having failed to achieve actuality, they were not doomed from the start to fail, and were in their time live options. The square box on the X branch indicating the node of the Tree of *u* and distinguishing the Tree of *u* from all the other sub-trees of *t* is making heavy drafts on contingent actuality, inappropriate in arguments about logical validity. It is just a matter of particular happenstance that the present emerged as it did, and we cannot use that fact to pick out one sub-tree from all the others. Semantic validity, as always, depends on the class of models being considered: a well-formed formula is semantically valid if it *be assigned* the value TRUE in every model of the relevant kind; if the range of model is altered, a well-formed formula that had hitherto *been assigned* TRUE in all of them, may now *be assigned* FALSE in some of them, and *vice versa*; so that if we restrict the class of models, we lose the generality we ought to seek.

(iii) Branches

McKim and Davis meet this line of criticism by arguing that no branch is being excluded, since the thesis

$$p_u \rightarrow P_uF_tp_u$$

holds good not just along the heavily marked branch representing the actual course of history leading up to its being white

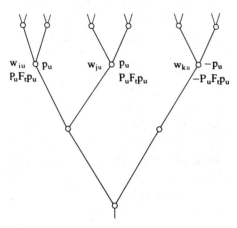

Figure 9:5.

now, but along every branch there is or could be. *Whatever* other possible branch, of the Tree of *t* we choose, it will have a possible world, w_{iu}, or w_{ju}, or w_{ku}, *etc.* at *u*, in which **either** p_u and $P_uF_tp_u$ **both** *be* True or else they **both** *be* False: for if in some branch including w_{ju}, which we may symbolize by W_j, p_u *be* True, as in the figure it is along the branches including w_{ju}, then along the branch leading from the node of the Tree of *t*, the argument goes exactly as before, so that along this branch $P_uF_tp_u$ *be* True also; and if p_u *be* False, as in the figure it is along the branches including w_{ku}, then $p_u \rightarrow P_uF_tp_u$ *be* True trivially: so that either way

$$p_u \rightarrow P_uF_tp_u$$

is satisfied along this branch; or, as we can put it,
$$W_j \vDash p_u \rightarrow P_uF_tp_u \text{ and } W_k \vDash p_u \rightarrow P_uF_tp_u.$$

So this holds either way, and thus is true irrespective of which branch we have chosen. So in every 'W*' branch of the Tree of t, since it contains either p_u or $-p_u$,

$$W_* \vDash p_u \rightarrow P_uF_tp_u;$$

and $\qquad W_* \vDash P_uF_tp_u \rightarrow p_u$ similarly.

Since the Tree of t $\mathbf{T}(t)$ is simply the union of all its branches, it seems that the two implications are therefore satisfied in $\mathbf{T}(t)$ as well, so constituting a reasonable proof of semantic validity. This, in effect, is the argument of McKim and Davis. In their criticism of McArthur they define a set of 'tau-models' (pp. 235–6) in terms not only of a set of moments, M, an accessibility relation, R, possible world histories, B, and a valuation function, V, but of a further function, f, which picks out, for any date, m, a particular branch, b, which could play the part of the actual future from date m onwards. Granted this, it is easy to extend the valuation function to cover the actual future $\mathbf{F}$, as well as the must-be future $\boxed{\mathbf{F}}$ and the may-be future $\langle\!\mathbf{F}\!\rangle$.

But should we grant this? The function, f, picks out a particular possible world history, b, and although we do not know which one it actually is, it seems to confer on it a determinate ontological status. A logical determinist would be quite ready to concede that we do not know which possible course of events will turn out to be the actual one, but would argue that whichever one it is, it narrows the range of possible future alternatives to just one that cannot but take place. McKim and Davis counter that the appearance of particularity and determinateness is misleading. We do not need to be able to specify which branch is *the* actual future branch, but only that it be *a* branch, since the claim to semantic validity is a claim that the well-formed formula in question holds in *all* models, so that the particular details of each model are irrelevant, so long as each is specified sufficiently to license the inference in question. It is a standard move. In traditional geometry we consider *a* triangle, a representative particular triangle, which we may draw on the board, with determinate angles and a determinate size, but its shape and size do not matter. We argue only from the features that are given as being relevant, and having reached a conclusion based on these features alone, argue further that since it holds of this triangle, it must hold of *any* triangle, and therefore of all triangles that satisfy the initial specification. Exactly similar modes of inference are encapsulated in the Rule of Generalisation in modern formal logic that licenses the inference from a well-formed formula containing a free variable or arbitrary name to the corresponding universally quantified one.

But still it might be objected that to use models based on branches, representing possible world histories, rather than on Trees themselves, was to beg an important ontological question. If to be is to be the value of a variable, then to quantify over branches rather than Trees is to suggest that branches are the stuff of which the future is made, and that underlying everything there is a set of definite possible world histories, just one of which, although *we* do not know which one, is *the* actual world history. It is like the Argument from Angels in chapter 3.[10] If once there exists a *determinate* body of angels who, without being infallible, are nevertheless always right, we begin to feel that their foretellings, though supposedly fallible, are none the less threatening our freedom. So, too, if there is a determinate branch which is the actual course of future events, then all the other possibilities fade away into unreality, and it is definite what the future will be, even though we cannot know it.[11]

The dependence on branches is likewise shown in Burgess's formulation of the 'Actualist' position.[12] Burgess defines semantic validity for Actualism by means of satisfaction in a Tree Frame, $(T, <)$, where T corresponds to our set of possible worlds (or states of affairs) at a time, and $<$ to Q^{-1}; and specifies not only x, the state of affairs at a time, but X, a branch containing x. Then given an A-valuation map, I, *assigning* to each (propositional) variable those pairs $\{x, X\}$ in which it *be* True, his crucial definition (p. 161) is[13]

$$T,I \vDash_{x,X} \text{FA} \quad \text{iff} \quad (\forall y \in T)(y \in X \ \& \ x < y \ \& \ T,I \vDash_{y,X} A)$$

---

[10] See above, ch. 3, § (vi), pp. 50–53.

[11] See Storrs McCall, 'The Strong Future Tense', *Notre Dame Journal of Formal Logic*, 20, 1979, pp. 489–90.

[12] Burgess, 'The Unreal Future', pp. 157–69.

[13] Epsilon stands for set-theoretical membership.

which is to say that FA is satisfied, with respect to *x and X*, iff there is a later world, *y*, on *X*, in which A is satisfied. This is the only place where the branch, *X*, comes into the argument – in his formulation of Antactualism, Burgess dispenses with the branches altogether, and uses only $T,I \vDash_y A$ etc., with no second subscript after the $\vDash$. Once, again, it would seem that if it be metaphysically illegitimate to postulate branches, or possible courses of world history, as sufficiently existent entities to quantify over, we cannot formulate a coherent semantics for the actual future, while if it be allowed that branches do in some sense exist, we are in danger of being tied down to just one branch, whatever it may be, and however little we can know what it will prove to be, with the consequence that all our future freedom of action is, really speaking although unbeknownst to us, fully foreclosed.

(iv) Sub-trees

We can meet this criticism, by considering sub-trees instead of branches. It seemed that to move the set of models from Trees to branches was to make a shift that might be metaphysically dubious: branches have a linearity about them that carries determinist overtones. But sub-trees are the same shape as Trees. They are linear in part – they have linear trunks – which will be enough for our purpose, but they are not altogether or embarrassingly linear. So we rephrase the argument we gave before, only this time in terms of sub-trees instead of branches. If it is white now, at time *u*, then the Tree of *u*, which can be regarded as a sub-tree of the Tree of *t*, is one in which every valuation *assign* the value TRUE to its being white at *u*. In this sub-tree it would be the case that every valuation at time *t* was one which also *assign* TRUE to p_u. So, **relative to this sub-tree**, (the one marked in Figure 9:6) $F_t p_u$ *be* True at *t*; i.e. $F_t p_u$ was, so far as this sub-tree go, temporally true at *t*. So, relative to this sub-tree, $P_u F_t p_u$ *be* True at *t*; that is to say, $P_u F_t p_u$ is, so far as this sub-tree go, temporally true at time *u*. But this is

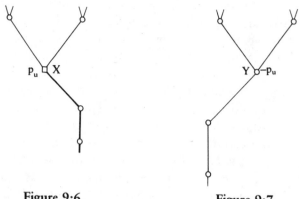

Figure 9:6.   **Figure 9:7.**

the actual sub-tree. Although there were other possible sub-trees of the Tree of t, each with nodes at u – e.g. that marked Y in Figure 9:7 – they are no longer relevant. The actual Tree is the sub-tree shown in the diagram above, and marked X in Figure 9:8. Hence $P_u F_t p_u$ is a semantic consequence of p_u in this model. But what about the other sub-trees? Although if we go back along the actual sub-tree, and come forward along it again, we shall, indeed, find in that future what has actually

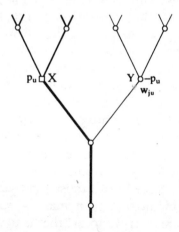

Figure 9:8.

taken place, by what right do we confine ourselves, when considering things from the temporal standpoint of time t, to the sub-tree that *actually* will turn out to be the relevant one at time u? If we had traced out some other sub-tree, say that marked Y, leading to a possible world, w_{ju} in which $-p_u$ *be* True, they would have been different. No doubt. But then although $\mathbf{P}_u\mathbf{F}_tp_u$ would not *be* True along the sub-tree leading to Y, neither would p_u, so that $p_u \rightarrow \mathbf{P}_u\mathbf{F}_tp_u$ would **still** *be* True. The argument is essentially a two-stage one. We confine ourselves to a certain range of models – the u sub-trees of the Tree of t, as we shall call them – which, one or other of them, must be the actual Tree at time u, and then timelessly generalise over them. It does not matter what our temporal standpoint actually is: we may well be at time t, and not know anything about the future course of events, but only that at time u the Tree of u, whichever it is, will be a sub-tree of the Tree of t, and will have its node at u. So we can take up that temporal standpoint in our imagination, and work out that, from that point of view, we shall be able to go back along a unique trunk, to time t, and return, again uniquely, to time u. This will be so whatever the actual u Tree may prove to be. We do not need to know *what* it is, but only *that* it will, then, be an actual Tree, with a node, at which **either** p_u *be* True, in which case $\mathbf{P}_u\mathbf{F}_tp_u$ *be* True also, so that by ordinary propositional calculus $p_u \rightarrow \mathbf{P}_u\mathbf{F}_tp_u$ *be* True, or p_u *be* False, in which case again by ordinary propositional calculus $p_u \rightarrow \mathbf{P}_u\mathbf{F}_tp_u$ *be* True too. We are using our ability to take up other temporal standpoints than the one we actually happen to occupy in order to define the relevant range of semantic models, namely the sub-trees of the the actual Tree of t, which are of the form the actual Tree at u must be, and therefore must have their node at u. These exhaust the range of relevant models for evaluating the temporal truth of propositions with respect to time u. So, if something holds in all such sub-trees, since one of those sub-trees must turn out to be the actual Tree at time u, it must be valid, whatever the course of events may prove to be, that is to say

$$\vDash p_u \rightarrow \mathbf{P}_u\mathbf{F}_tp_u.$$

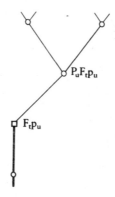

Figure 9:9.

Exactly the same argument holds for the converse implication. The only way for $P_uF_tp_u$ to *be* True in the u sub-tree is for p_u to *be* True also. Hence p_u is equally much a semantic consequence of $P_uF_tp_u$ in the relevant model. And so, generally, the well-formed formula $P_uF_tp_u \rightarrow p_u$ holds good in each and every relevant model, and so is semantically valid. We can combine these two results and say:

$$\vDash p_u \leftrightarrow P_uF_tp_u$$

(v) The Sea Battle

In every possible u sub-tree the sea-battle argument holds also. Either there will be a sea battle, or there will not. For in any sub-tree with its node at u, there is only one Q-accessible world of date u, in which p_u must *have* one uniform truth-value, either TRUE or else FALSE. If it *be* TRUE (shown in Figure 9:10.), then for any $t < u$, F_tp_u *be* also *assigned* the truth-value TRUE in that sub-tree, and hence, $F_tp_u \vee F_t-p_u$ too: if it *be* FALSE (shown in Figure 9:11.), then $-p_u$ *be assigned* the truth-value TRUE and so also F_t-p_u likewise, and hence $F_tp_u \vee F_t-p_u$ *be assigned* the truth-value TRUE: so either way $F_tp_u \vee F_t-p_u$ *be assigned* the value TRUE. So the sub-tree in question *make* the

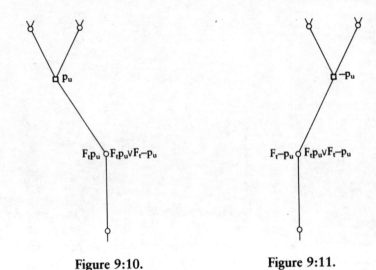

Figure 9:10.   Figure 9:11.

well-formed formula $F_t p_u \lor F_t -p_u$ come out True. Since this holds of any possible sub-tree with node u, it holds of every, and

$$F_t p_u \lor F_t -p_u$$

be therefore True for all possible sub-trees with node u of the t Tree, and so is valid. Hence, we argue, it is valid for the t Tree itself. For although we do not know, cannot specify, and indeed believe it is unspecifiable, *which* possible sub-tree with mode u will become *the* actual Tree of u, we do know what it is *like*. We do not know whether it *bring* out p_u as True or as False, but we do know that since its node is at time u, there is only one possible world at that time on that particular sub-tree, so that p_u must *have* the same truth-value in every world of that date in that sub-tree, and so either *make* p_u True or else *make* $-p_u$ True throughout that sub-tree. We do not know whether there will be a sea battle tomorrow or not, but we do know that tomorrow when it comes must be either a sea-battle day or not a sea-battle day. And then we can generalise again, because it does not matter which t or u we took in our example – any t or u would do; whatever day tomorrow happens to be, either it will be a sea-battle tomorrow or else it will not: and whatever day today happens to be, either there will be a sea battle tomorrow or else there will not. And so we say that $F_t p_u \vee F_t - p_u$ is valid altogether, that is,

$$\vdash F_t p_u \vee F_t - p_u.$$

(vi) Modal Vacuity of the Actual

Two inter-derivable theses, expressing Commutativity with Negation and Division under Disjunction, characterize modal vacuity.[14] In the case of the actual future operator, F_t, they are:

$$-F_t p_u \rightarrow F_t - p_u$$

and

$$F_t(p_u \vee q_u) \rightarrow (F_t p_u \vee F_t q_u).$$

The former is simply a propositional calculus variant on the sea-battle thesis already shown to be valid. The latter follows from the uniqueness of the node by an argument similar to that given for the sea battle above. Since there is only one state of affairs which *obtain* at the node of any possible sub-tree, $p_u \vee q_u$ can *be* True only if either p_u *be* True or q_u *be* True: if the former $F_t p_u$ *be* True in the same possible sub-tree; if the latter $F_t q_u$: and so either way, $F_t p_u \vee F_t q_u$ *be* True in that possible sub-tree. While if neither p_u nor q_u *be* True, then $F_t(p_u \vee q_u)$ *be* not True either, so that

$$\vdash F_t(p_u \vee q_u) \rightarrow F_t p_u \vee F_t q_u$$

be True in the possible sub-tree with node at u under consideration. Since it holds of any sub-tree, it holds of every sub-tree, and once again we find that the thesis is valid for all possible sub-trees with node at u.

Modal vacuity is represented semantically by the uniqueness of the accessible world. If there is only one accessible world,

[14] See above, § (ii), p. 165, and ch. 5, § (i), p. 89.

then either all of them possess a certain feature, e.g. that of containing p_u, or else none of them do: there is no third alternative of some, but not all, of the accessible worlds possessing the feature in question, thereby giving rise to some indeterminacy of truth-value. Tree Semantics secures the modal vacuity of the present and the past by reason of the trunk of the Tree being unique: sub-tree semantics is being called in aid to provide a comparable uniqueness and secure a similar vacuity for the actual future.

The essential difference between the actual future operator F_t and the other operators is that F_t requires us to use possible sub-trees with their nodes at a future date – that of the well-formed formula governed by F_t – whereas all the other operators are valid in Trees of the date indicated by the index. For the other operators and for temporal truth generally, it is a determinate question whether truth conditions are satisfied or not. A given actual Tree picks out a certain number of states of affairs at the relevant date – those on the Tree – as relevant, and the valuation of complex well-formed formulae proceeds in a straightforward manner. With F_t, however, the valuation is relative to future possible sub-trees. Given one possible sub-tree, $F_t p_u$ would *be* True, given another it would *be* False. Nevertheless, some propositions *come out* True, whatever (appropriately dated) future possible sub-tree is chosen, because although it is not fully determinate which sub-tree *be* the actual one, it is fixed what state of affairs it characterizes at time t, and also what its shape *be* – essentially that it *be*, at the node, unique, 'containing' only one state of affairs, that is to say, characterizing one state of affairs completely and determinately.

(vii) Temporal Truth

Sub-tree semantics furnishes us with a very similar account of temporal truth for statements about the future. We first argue formally. We have already established

$$\vDash p_\mathrm{u} \to \mathbf{P_u F}_t p_\mathrm{u}.^{15}$$

From this it follows that if p_u be temporally true at time u or later, then $\mathbf{P_u F}_t p_\mathrm{u}$ be temporally true then too. But if $\mathbf{P_u F}_t p_\mathrm{u}$ be temporally true at time u, $\mathbf{F}_t p_\mathrm{u}$ be temporally true too, because $\mathbf{P_u}$, being a special case of $\Box_t$, is T-like.[16] Conversely, if p_u be temporally false at time u, then $\mathbf{F}_t p_\mathrm{u}$ must be temporally false too. So $\mathbf{F}_t p_\mathrm{u}$ is temporally true iff p_u is temporally true at time u. That is, although the date of $\mathbf{F}_t p_\mathrm{u}$ is explicitly that of its subscript t, its temporal truth is that of the well-formed formula governed by $\mathbf{F}_t$, namely u.

The underlying reason lies in our sub-tree semantics. We want to ascribe temporal truth to $\mathbf{F}_t p_\mathrm{u}$. In order to do so, we have to consider not the Tree of t, but some sub-tree with node at u. $\mathbf{F}_t p_\mathrm{u}$ *be* True only with respect to some sub-tree with node at u, and in such a sub-tree *be* True iff p_u *be* True in that sub-tree too. In such a sub-tree p_u *be* tenselessly True, iff p_u is temporally true at time u. So $\mathbf{F}_t p_\mathrm{u}$ be True iff p_u is temporally true. In general p_u is temporally true only at time u and thereafter. So the condition for $\mathbf{F}_t p_\mathrm{u}$ being temporally true is one that is satisfied only at time u and thereafter. The actual future, which has the Reichenbachian structure S–R,E instead of S,R–E, can have temporal truth, but a temporal truth that is dated to the u – the time of the event – rather than to the t – the time of utterance. Although $\mathbf{F}_t p_\mathrm{u}$ can be uttered, and discussed, at time t, in advance of the event, its temporal truth is, unlike the apparently similar $\boxed{\mathrm{F}}_t p_\mathrm{u}$, to be dated to u. Valedictory truth is the appropriate truth for actual futures.

(viii) Defeasible Truth

The defeasible ascription of indefectible truth to predictions involves two modalities. Indefectibility is a temporal necessity. The condition under which a statement in the posterior present

---

[15] See above, end of § (iv), p. 172.
[16] See above, ch. 6, § (i), p. 113.

(S,R–E) tense can properly be made at time t is that the future is already present in its causes, so that, whatever other variation there may be in the course of events, the event predicted is part of every possible world w_{ir} of the relevant date r, in Tree of t $T(t)$, that is to say the event has got to occur.[17] But the modality of the 'got to' is different from that implicit in the possibility of defeat by the subsequent turn of events. The admiral may have given the go-ahead for a sea battle tomorrow, and then have a last-minute change of mind, or be recalled to Athens, or be prevented from engaging by a sudden storm. These possibilities are wider than those taken into account by subordinate officers, who properly predicted, on the strength of the admiral's order, that there was going to be a sea battle, and instructed the rowers to grease their rowlocks accordingly. The different modalities give rise to different trees, the more stringent modalities giving rise to wider ranges of possibility, and hence to more branching.[18] If we take into account the perpetual possibility of a man's changing his mind up until the deed is actually done, there are many, many possible future world histories in which predictions based on the assumption that everyone acts morally, reasonably and sensibly, do not come true. Even with less wide-ranging defeaters than the universal *homine volente* clause, there are some possible courses of events in which the prediction proves false. Only when the moment of truth has come can we be absolutely sure that our properly grounded predictions will not be defeated by the turn of events.

We therefore have two trees, based on two accessibility relations, Q_1 and Q_2, representing two modalities, the weaker one on which the prediction was properly grounded, the stronger one according to which it might yet be defeated. If we superimpose them,[19] we have something like the diagram portrayed. In the tree with fewer branches, marked in more heavily, all the possible courses of events have a sea battle

[17] See above ch. 8, § (v), p. 153.

[18] See above, ch. 8, § (i), p. 138.

[19] We here discount the case noted in ch. 8, § (i), p. 138, that neither tree can be included in the other.

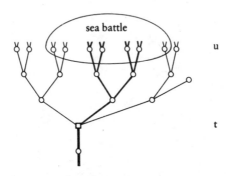

Figure 9:12.

occurring tomorrow: whereas the more branched tree, allowing perhaps for a change of mind on the part of the admiral or the Athenian *demos*, or for all meteorologically possible storms, has many possible worlds tomorrow in which no sea battle takes place.

Since we are only concerned with tomorrow's sea battle, we can without loss of generality project all the different possible future courses of events in the smaller, Q_1, tree onto a single branch, neglecting the variations in which the pre-battle chariot races are won by Antilochus or by Ajax. We then have the same diagram when we consider the wider, Q_2, tree as for the actual future. Among all the possible branches one is singled out as that which will actually be realised. According to the wider, Q_2-modality, there is no 'got to' about it, no reason for saying that it is going to. It is not a prediction, only a conjecture, which would admit of only valedictory truth, ostensibly antedated to t, the time of its utterance S, but really to be dated to u, the moment of truth when it could no longer prove wrong. As we change the modal focus, so we change the tree semantics and sort of truth ascribed. So long as we are operating with the less stringent modality, Q_1, with a correspondingly narrower range of possibility, we predict, claiming indefectible truth for our prediction of an event that is going to happen since it is already present in its causes: but when we widen our horizon and

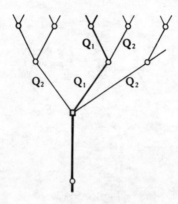

Figure 9:13.

operate with the more stringent modality, Q_2, the necessity disappears, and we no longer claim indefectible truth for our prediction, but merely conjecture that the event will actually take place, in which case our conjecture will have turned out true.

For the most part, however, we operate bifocally in a mixed mode of discourse. Predominantly we predict, ascribing indefectible truth, on the basis of the Q_1-tree, and dating it to the time of the prediction: but then we hedge, and acknowledge that there is many a slip between the cup and the lip, and acknowledge a defeasibility which can itself be conclusively defeated only by the passage of time, so that only *ex post facto* can we be absolutely sure that our Q_1-prediction would not be defeated by the turn of events, and was Q_2-valedictorily true.

(ix) Conclusion

The criticisms levelled against McKim and Davis can be met. They have shown that the semantic validity of a well-formed formula can be established if each of an appropriate range of models satisfies it, even though we cannot specify any of the models completely or say which it is we should really be considering. But the appropriate range of models is not the branches of the Tree at t, but the 'u sub-trees' of the Tree at t. These carry with them no suggestion of a linear future, but only of a determinate present and a linear past. And we can reasonably regard them as conceivable, and so suitable, entities to quantify over, since we can imagine ourselves at some other date, either future or past, than the present. I can imagine myself tomorrow, surveying the sea off Naupactus, and although I do not now know whether or not there will be a sea battle, I can work out from what it would be like to be tomorrow that then it will be a determinate matter whether or not it is true that there is a sea battle. And equally then, if there is a sea battle, I can valedictorily affirm that the simple future $F_{28}p_{29}$ has turned out to be true, and can with hindsight say that if anyone had said 'There will be a sea battle on February 29th, 2000', he would have spoken truly.

With predictions in the posterior present tense (S,R–E) the conditions for truth are more stringent, and if satisfied, the temporal truth is properly dated to the Reichenbachian reference point R, which is contemporaneous with the date of utterance S, rather than that of the event. But in this uncertain world it is a big IF. However carefully I assess the evidence, and however well warranted my prediction is, there is always, until the event has actually occurred, some possibility in some modality that things may go awry, and that my prediction be falsified in the event. And so, although we rightly ascribe to a well-warranted prediction that did not prove mistaken a temporal indefectible truth from the time it was made, we keep our fingers crossed while there remains any possibility of its not coming true, and relax only when we can safely assert that it was valedictorily true that our defeasible ascription of indefectible truth would not actually be defeated in the event.

10

A Modal Derivation of Time

(i) Time and Modality

Thus far in this book I have attempted to explain our concepts of temporal truth and temporal modality, taking for granted certain properties of time. I have assumed that time is linear, not cyclic, that there are alternative Q-possible courses of events open to us in the future, and that the past is unalterable, and so in our representation trunk-like. I have tacitly assumed that time is at least dense, if not continuous in the mathematicians' sense, and that it has no end, and that some timeless truths of logic and omnitemporal truths of physics can be said to be true from the beginning of time, if time has a beginning, and in any case *ante omnia saecula*. These are natural assumptions, which are not likely to be seriously contested, except perhaps the branchingness towards the future of possible courses of events. But although they seem to be true, it is not clear why they should be true. If it were possible to give a modal derivation of time, and account for some of the features of time as consequences of its modal origins, it would go far to making time, if not less mysterious, at least more intelligible. We have some idea that the future is open, the present actual, and the past unalterable, and we think of time as being the passage from possibility through actuality to necessity. Can we from this and other similar insights derive the structure of time that we commonly take for granted?

It is difficult to derive time, because we take it so much for granted that we cannot divest ourselves of temporal language and temporal assumptions. In our very symbolism for propositions and propositional contents we have assumed some temporal differentiation in the subscripts, p_t, p_u, *etc.*, and indeed, if once we admit that things, or states of affairs, or possible worlds, may be qualitatively identical but numerically distinct, we are committed to there being some differentiating parameter which may well be a temporal one.[1] But for the present we do not avail ourselves of arguments from numerical and qualitative difference, and drop indices, writing simply, *p*, *q*, *r*. We are not reverting to Prior's usage. For him, *p*, *q*, *r* are temporally indefinite propositions, and **P***p* means that *p* was the case sometime. Here, however, *p* will not be temporally indefinite, waiting to be bound by, as it were, a quantifier, *sometime* or *always*: it is complete in itself, but carries no temporal indicator. This understanding of *p*, *q*, *r*, etc. does not rule out the use of Prior's tense operators, but encourages their being understood in a more dynamic way, giving not a static picture of how different temporal standpoints, R, may be related to one another, but a dynamic one of how different moments of utterance, S, must be. Of course, we shall need to distinguish standpoints from one another, since they would be otherwise indistinguishable, and likewise occasions of utterance. They will have to be indexed. But to avoid any appearance of begging the question, we shall call them modal standpoints rather than temporal standpoints, and shall use *i*, *j*, *k*, *l*, as subscript indices instead of *s*, *t*, *u*, *v*.

Different modal logics are characterized by different relational structures. If there are arguments for time having the modal structure it has, they constitute arguments also for its having the relational, and perhaps even the topological, structure we find ourselves ascribing to it. If we can start with a family of modal operators, indexed by some set, not necessarily linear or dense or continuous, we may hope then to argue that

[1] See J. R. Lucas, *A Treatise on Time and Space*, London, 1973, §§17–26, pp. 99–123; and R. C. S. Walker, *Kant*, London, 1978, esp. chs. 3 and 4, pp. 28–59.

the indices must be linearly ordered, with a tree-like structure for possible worlds and possible standpoints.

(ii) Transitivity of the Future

Future possibility is S4-like. What is possibly possible is possible. If it may be that there may be a sea battle, then there may be a sea battle. If I am an agent choosing, and I can choose to be in a position to choose, I can choose. If there is some probability between 0 and 1 of there being some probability between 0 and 1 of a particular event's happening, then there is some probability between 0 and 1 of its happening. For this modality then the thesis $\Diamond\Diamond p \to \Diamond p$ will obtain, which is equivalent to $\Box p \to \Box\Box p$, the axiom 4, characteristic of S4. And then, as we have seen,[2] the underlying accessibility relation Q must be transitive, in which case Q^{-1} must be transitive too, and $\Box^{-1}$ must satisfy axiom 4 too.

(iii) Linearity of the Past

But we want more. The modal logic of the past operator contains 4, the typical thesis of S4, but is stronger – something more like S4.3. For although there are alternative possible courses of events towards the future, there are none towards the past. Linearity is expressed by Prior in the law of trichotomy:

$$\mathbf{P}p \ \& \ \mathbf{P}q \to (\mathbf{P}(p\&q) \lor \mathbf{P}(p\&\mathbf{P}q) \lor \mathbf{P}(\mathbf{P}p\&q))$$

if p is past and q is past then either $p\&q$ is past, *i.e.* there was a time before now when $p\&q$ was true, or $p\&\mathbf{P}q$ is past, *i.e.* there was a time before now when p was true and q was past,

---

[2] See above, ch. 5, § (iii), pp. 92–3.

or Pp&q is past, *i.e.* there was a time before now when q was true and p was past. An alternative, perhaps more intuitive, formulation uses mixed operators: it runs

$$\text{FP}p \rightarrow (p \vee \text{P}p \vee \text{F}p)$$

if something will have been the case, then either it is the case now or it already has happened or it will subsequently be the case. We need to ask whether we have any reason for positing the modal analogues to these,

$$\Diamond^{-1}p \;\&\; \Diamond^{-1}q \rightarrow \quad\quad (\Diamond^{-1}(p\&q) \vee \Diamond^{-1}(p\&\Diamond^{-1}q) \vee \\ \Diamond^{-1}(\Diamond^{-1}p\&q))$$

or

$$\Diamond\Diamond^{-1}p \rightarrow (p \vee \Diamond^{-1}p \vee \Diamond p)$$

Why should Q^{-1} be linear, that is to say, why should the past be unalterable?[3] There are three arguments, one from high theology, one from solid metaphysics, and one as a condition of responsible agency. We can best appreciate the force of these by supposing that the past might not be linear, and that the modality of the past operator might be **S4.2** instead of **S4.3**. Prior quotes Lukasiewicz:

If, of the future, only that part is real today which is causally determined by the present time; . . . then also, of the past, only that part is real today which is still active today in its effects. Facts whose effects are wholly exhausted, so that even an omniscient mind could not infer them from facts happening today, belong to the realm of possibility. We cannot say of them that they *were* but only that they were *possible*. And this is as well. In the life of each of us there occur grievous times of suffering and even more grievous times of guilt. We should be glad to wipe out these times not only from our memories but from reality. Now we are at liberty to believe that when all the

---

[3] We need to distinguish the actual linearity of Q^{-1} from the linearity of the inverse of the relation Q^*, which we can, thanks to Szpilrajn's theorem, construct. This, as we saw in ch. 5, § (iv), p. 99, expresses the linearity of past dates rather than the unalterability of the past.

consequences of those fatal times are exhausted, even if this happened only *after* our death, then they too will be erased from the world of reality and pass over to the domain of possibility.[4]

Prior does not endorse this way of talking, arguing that even if it is all the same in a hundred years time, there will still be the difference that one thing rather than another *has been* the case. But simply to make that assertion is to beg the question. If, however, there were an omniscient mind, not simply a Laplacian intelligence, capable of retrodicting, no matter how complicated the calculations, from presently available data, but able to remember as well, then the past cannot be wiped out, because even if it has exhausted its causal effect, and no longer has any consequences in the present state of the world, it will still exist in the unforgetful memory of God. No theist can deny the unalterability of the past. Just as what we do in the present always matters, because what we do is known to God, and He cares, so what we have done always remains, because He remembers. It may be forgiven, but it cannot be forgotten. If there is a God who remembers everything that happens, then anything that has happened is remembered by God, and cannot be other than as He remembers it, and so cannot be other than it was. A theist is thus committed to the unalterability of the past, and so to its having the form of a linear trunk, not a branching set of roots. Although Agathon's verses,

μόνου γὰρ αὐτοῦ καὶ θεὸς στερίσκεται,
ἀγένητα ποιεῖν ἄσσ' ἂν ᾖ πεπραγμένα

Of this is even God deprived – The power to make undone what has already been done,[5]

---

[4] Jan Lukasiewicz, 'On Determinism', in L.Borkowski, ed. *Selected Works*, North Holland, 1970, pp. 127–8; reprinted in Storrs MacCall, ed., *Polish Logic*, Oxford, 1967, pp. 38–9; quoted by A. N. Prior, *Past, Present and Future*, Oxford, 1967, p. 28.

[5] Quoted by Aristotle, *Nicomachean Ethics*, VI, 2, 6, 1139^{b}10, 11. See also Aquinas, *Summa contra Gentiles*, II, 25, 1023, *Deus non potest facere quod praeteritum non fuerit*, God cannot make the past not to have been.

portray the unalterability of the past as a limitation on the power of God, it should be seen not so much as a limitation as a concomitant of His being a person.

We can consider a secular analogue in Broad's principle of accretion, whereby the deposit of true facts increases from one generation to another, from one moment to another, or as we should now want to say, from one modal standpoint to another.[6] If there is some principle of the set of temporal truths being monotonically ordered by the set-inclusion relation, then the accretion of truth will correspond to the passage of time, which will be linearly ordered by the monotonic increase in the deposit of truth.

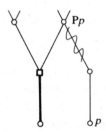

Figure 10:1.

The alternative condition for linearity, involving mixed modalities, can be expressed, using Prior's tense operators

$$\mathbf{F}\mathbf{P}p \rightarrow (p \vee \mathbf{P}p \vee \mathbf{F}p)$$

and we may ask whether this condition could be justified by non-temporal argument. In its ordinary temporal interpretation it limits the range of the future perfect to what is or was or will be the case, excluding the possibility that at some future juncture a new past, hitherto unknown to us, may become accessible. Our possible achievements are limited to what we do or have done or may yet do. And this seems appropriate for the concept of agency too, divorced, so far as it can be, from adventitious tinctures of temporality. An agent concerned not

---

[6] C. D. Broad, *Scientific Thought*, London, 1923, p. 73.

just with what he can do but with what he can bring it about that it be unalterably the case needs to survey only what is accessible to him: he is in a position, modally speaking, to know what can be unalterable; it is simply either what is already unalterable or what he is doing or what can still be done.

The principle that there is nothing 'round the corner' modally speaking is an attractive one. Although Lukasiewicz could find comfort in thinking it would all be the same in a hundred years time, whatever he did, many dissidents in later years have been impelled to stand up and dissociate themselves from what was being done, and to make it clear that they were not consenting to it. They know their actions will have no effect on the course of events, but it matters to them all the same that they should be witnesses to the values they hold dear. It would be a monstrous injustice if they were subsequently held to be responsible for what they had denounced, on the grounds that the passage of time had merged their actions with those they had condemned. Hence the need to rule out the possibility of Pp by virtue, not of something accessible to or from the agent, but of some other backward link. All possibility of of achievement is being funnelled, so to speak, through the agent, without any possibility of adventitious achievement. He is on his own. It is up to him. He is master of his fate, so far as anyone can be – though, of course, the success of his endeavours may very much depend on factors outside his control. But it cannot become the case that *he* has done something, without his having done it already, or doing it, or being able to do it subsequently. He will be responsible for what he will have done, but this is only what he has done or does then or will do subsequently.

Some care is necessary in working out the details. It has been formulated in terms of Priorean tense operators, with their indefinite date reference, and we need to distinguish the sense in which what was sometimes the case is unalterable and necessary from the more common association between □p and Hp, what has always been the case. Our principle concerns the interplay between $\Diamond$ and $\Diamond^{-1}$: whenever we have a proposition p viewed from a modal standpoint so that $\Diamond^{-1}p$ holds, which is such that when viewed from another modal standpoint it is appropriate to say $\Diamond\Diamond^{-1}p$, then it follows that either p or $\Diamond p$ or $\Diamond^{-1}p$.

(iv) Non-linearity of the Future

The question will naturally be asked why the converse condition should not be required too: if we insist that

$$\mathbf{F}\mathbf{P}p \rightarrow (p \vee \mathbf{P}p \vee \mathbf{F}p)$$

why should we not insist also that

$$\mathbf{P}\mathbf{F}p \rightarrow (p \vee \mathbf{P}p \vee \mathbf{F}p)?$$

The answer lies in the irrevocable exclusivity of choice. To choose is to exclude: if I choose to marry Jane, I forgo the possibility of marrying Jean. What was open from one modal standpoint cannot be open from another where the choice has been made, and some possibilities have been actualised and others rejected. A similar point can be seen if we consider probability as the numerical filling out of possibility.[7] If we consider the probability of rain tomorrow, February 29th, 2000, it will vary, as the clouds gather and disperse, the cold front comes in from Russia and the depression approaches from Iceland, sometimes getting larger, sometimes smaller. But if once it becomes zero, it remains zero thereafter. If we work it out in the calculus of probabilities, with $\mathrm{prob}_t(x_v)$ being the probability at time t that x_v will occur at time v, then

$$\mathrm{prob}_t(x_v) = \Sigma(\mathrm{prob}_t(y_{iu}) \times \mathrm{prob}_u(x/y_{iu}))$$

where y_{iu} are the states of affairs at an intermediate time between t and the date of x_v, and $\mathrm{prob}_u(x_v/y_{iu})$ are the conditional probabilities of x given y_{iu}: If ever $\mathrm{prob}_t(x) = 0$, it can be only because each of the non-negative components of

---

[7] See above, ch. 4, § (iv), p. 76, and ch. 5, § (ii), pp. 90–93.

$\Sigma(\text{prob}_t(y_{iu}) \times \text{prob}_u(x/y_{iu}))$ is equal to 0, which in turn can be only because in a component either $\text{prob}_t(y_{iu})$ or $\text{prob}_u(x/y_{iu})$ is 0. If the former, there is no chance of the state of affairs y_{iu} coming to pass; if the latter, there is no chance of x coming to pass if y_{iu} does. So either way, if once the probability becomes 0, it remains 0 thereafter. A similar argument holds for 1, by a similar train of reasoning about not-x. Probabilities can fluctuate with time within the open interval (0,1), but if once they reach either boundary, they stick there. There is thus built into the calculus of probabilities the possibility of 'losing' branches as various probabilities mature into definite truth-values, 0 or 1. This is a temporal argument from the passage of time. If there is any real passage of time, then it is marked by the collapse of probabilities into certainties, and the foreclosing of possibilities that once were open, but are so no longer. In which case the future is open, and we have good reason not to stipulate the condition, expressed in modal terms

$$\Diamond^{-1}\Diamond p \to (p \vee \Diamond^{-1}p \vee \Diamond p).$$

The built-in asymmetry between $\Diamond$ and $\Diamond^{-1}$, or equivalently between $\square$ and $\square^{-1}$, corresponds to a difference between Q and Q^{-1}, that, while both are transitive, the former is not, and the latter is, linear. These two conditions together require that their relational structure be a Tree.

(v) Homogeneity and Uniqueness

Prior's tense logic gives a frozen picture from one point of view, and we are drawn to it because it seems to express the uniqueness and importance of *now*. But there are many potential *nows*. Each past instant was in its time Now, and every future instant will be. These two requirements can be expressed in quite general non-temporal terms. Every modal standpoint must be, in some sense, unique, and yet all must be on a par, in something of the same way as each person is a unique centre of consciousness and initiator of action, but is, in that, like every one else. As soon as we put together these two requirements on a partially ordered structure, we are led to a Tree. A Tree has a unique node, yet every point on a Tree is a node of some other Tree, either a sub-tree, or a Tree that itself has the given Tree as a sub-tree of itself.

Trees are the most general structures generally satisfying these requirements. Although a well-trimmed thicket, as in the diagram, has a unique node, x, from which indeed it would follow that

$$\mathbf{F}\mathbf{P}p \to (p \text{ v } \mathbf{P}p \text{ v } \mathbf{F}p)$$

this would not hold good of other points on the thicket, such as Y. If not only is the actual present unique but every past instant was in its time the actual present, then backward branching is ruled out, not only for the branches extending forwards from x, but for every branch whatever; so, if the present is unique the past must be unique too.

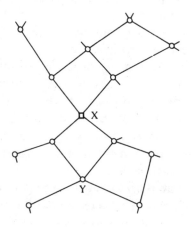

Figure 10:2.

A linear ordering satisfies the requirement of homogeneity, but not that of uniqueness. Every point on a linear ordering is *exactly* the same, so far as structure is concerned, as every other one, and so there is nothing to pick out the present as unique.

The present moment is no different from past and future moments. This is one reason why we feel that McTaggart's B-series fails to do justice to our intuitive notion of time, in that it represents only the relations of *earlier* and *later*, and does nothing to indicate the special status of the present moment as peculiarly momentous, and quite different from all others.[8] A linear series also fails to represent the anisotropy of time. *Earlier* and *later* not only are converse relations, but being mirror images of each other in a linear ordering, are exactly similar; whereas in our ordinary understanding of time *past* and *future* are not. A tree structure distinguishes Q from Q^{-1}, in that the former is one-many, whereas the latter is many-one. Q^{-1} gives rise to a function, assigning to each point on a Tree a determinate past, whereas Q does not give rise to a determinate function at all, representing the fact that the future is not determinate.

These last are, once again, temporal arguments. But the argument from uniqueness and generality is not, and once we have the thesis that possibly possibly *p* implies possibly *p*, we are committed to an accessibility relation which is transitive, and which, if it is not to be an equivalence relation, must be an ordering. There are many possible orderings, but only one satisfies the twin requirements of homogeneity and uniqueness. These are reasonable requirements. Each agent is unique. He has his own view of the world, and can make up his mind for himself. But each is rational, subject to some condition of universalisability, recognising others as rational agents like himself. Time must reflect the unique particularity of the existential moment at which I act, but also my ability to stand back from myself and be a spectator of all time. To stress either to the exclusion of the other is to have a defective concept. To live only in the present moment immersed in the immediacy of experience and reacting to it without forethought or reflection is to be submerged by particularity with no part in rational

[8] See above, ch. 1, pp. 9–10; see also J. E. McTaggart, 'The Unreality of Time', *Mind*, 18, 1908, pp. 457–84.

decision-making. Many people live like this, but their lives seem to them to have no significance in their own eyes, and come to nothing in the eyes of others. But equally to contemplate merely the one-dimensional temporal uniformity of the scientist is so much to take up the spectator's view as to preclude action and experience altogether. Time is not merely a homogeneous continuum, as I am not merely a unit of agency. Just as I am an individual who must act on my own responsibility now in the time of my mortal existence, so time is particularised by an egocentric particular, *now*, which is peculiarly the moment of action, though like many, many other moments which are potentially moments of action too.

(vi) The Maturation of Possibility

The argument so far has given some grounds for the difference between the future and the past, the uniqueness of the present, and hence for the tree-like structure of temporal standpoints. But it has only hinted at the passage of time. We have said that the future is open, the present actual and unique, and the past unalterable, but we have not really made good our claim that time is the passage from potentiality through actuality to unalterable necessity. Yet there are some intimations of actuality and passage in our normal notions of potentiality.

Probabilities imply time. If we ascribe a probability to a thing or an event, we are making a guarded prediction of its manifesting a particular property or of its occurring. Our ascription of probability would not have any content except under the supposition that the thing would in due course either have the property in question or not have the property in question, or that the event would either occur or not occur. If I say that the next throw of this coin will come down heads, then I am presupposing that the next throw is after the time of speaking. If I say that a particular radium atom has a 50 per cent probability of disintegrating in the next 1,700 years, then I am envisaging a subsequent situation in which either the radium atom has disintegrated – in which case it clearly is different from what it is now – or it has survived 1,700 years without disintegrating, and so is 1,700 years older than it is now. Every ascription of probability says less than a definite assertion, either that the thing does have, or that it does not have, the property in question, and says less than a definite prediction, either of the occurrence or of the non-occurrence, of an event. Probability has not yet given way to certainty. But sooner or later it must. Else, it has no point. Only if I do not know already what the outcome was or is or is going to be, is it worth making a probability judgement: I should be less than frank if I told you that the probability was such and such, when I actually knew for certain what was in fact the case. But only if there is an outcome to which my judgement of probability is a guide, albeit a guarded guide, is there any point in my uttering a probability judgement at all. There is an essential tension in all probability judgements between the partial and the complete. Probability judgements are, by definition, partial – they give way to certainty at 100 per cent and 0 per cent: but they get their meaning and their point from there being certain judgements, which although not as yet available will be available in due course. It follows, then, that for a probability to *be* between 0 and 1, it must also *be* either 0 or 1 and hence, if we are to avoid contradiction, it must be 0 or 1 somehow else. Some parameter is needed to enable us to say, without contradicting ourselves, that the outcome of the next toss is, with respect to one value of the parameter, indeterminate – say, 50 per cent heads –, but, with respect to another, determinate – say, heads. We cannot say in the same logical breath both say that it *be* (tenselessly) heads and not *be* heads, unless there *be* some parameter or some respect to distinguish the two.

But it is not obvious that the parameter has to be a temporal one. There is some difference, granted, perhaps some change, but not necessarily, at least so far as has been shown, a temporal change. Probabilities, like temperatures and other variables, can vary with respect to space as well as to time. The poker can be both hot and cold, not only by being hot today and cold tomorrow, but by being hot at one end and cold at the other. And thus an objector might doubt whether the difference between the partial knowledge given by a probability-judgement and the complete knowledge given by a definite judgement has to be a temporal one. The assumption, made by me and many others, that the parameter we have to invoke if experience is to be intelligible, must be a temporal one, has been called in question by Ralph Walker. And thus, though there is some argument for there being some parameter indexing those propositions to which probabilities are attached, so that the proposition can also be non-probabilistically asserted somewhere or somehow else, we have not yet given adequate reason for excluding the 'where', or indeed for insisting on its being spatial in a strict rather than a analogical sense of the word.

Mellor argues that in such a case we should say that we were dealing with different parts of the thing in question, whereas at different times it is the same thing.[9] This is certainly true as a matter of linguistic usage, but Walker constructs some ingenious examples to show that nevertheless identity can be construed in terms of a spatial rather than a temporal parameter, and that therefore some concept of identity can be given that does not presuppose that of time, and hence in countering Mellor's argument from linguistic usage.[10] Walker also succeeds in defending the idea of non-temporal experience against one argument of Strawson's, but does not succeed in making it free of all objection.[11]

The relevant objection for our present argument is that the maturation of possibilities is directed, whereas spatial and

[9] D. H. Mellor, *Real Time*, Cambridge, 1981.
[10] R. C. S. Walker, *Kant*, London, 1978, ch. 3, part 3, pp. 34–41.
[11] P. F. Strawson, *Individuals*, London, 1959. Many, myself included, find the idea of a non-temporal experience inconceivable. My claim, in J. R. Lucas, *A Treatise on Time and Space*, London, 1973, §2, p. 13, that time is a concomitant of consciousness, is of course only a claim, and I have been unable to argue for it, except by citing poetry. But other philosophers too have been unable to understand the suggestion of non-temporal experience, and though the argument from inability is never very strong, it may, if the inability is sufficiently widespread, carry some weight. But arguments would be better.

quasi-spatial parameters are not. Although Walker, in order to meet the next objection, introduces an anisotropic – that is, directed – spatial dimension by postulate, it is an unsatisfactory artifice. What is characteristic of space is isotropy, not anisotropy, and in any case the anisotropy, as we shall shortly see, is inadequately grounded.

A much stronger argument for temporality is the argument from action. Walker attempts to accommodate agency by postulating an anisotropic spatial dimension along which there would be analogues of memory and intention.[12] But fearing that this would be simply to re-introduce time under another name, he attempts to characterize agency in terms of counterfactuals.

... the idea of the will as efficacious, and as not determined by external factors, can be carried over quite successfully into the timeless world. For although the timeless subject cannot make any future decisions or perform any future actions, the world may still depend in various ways upon his will, in that if he had willed otherwise things would have been different. And his choice to will in one way rather than another may be free, in that there are no true counterfactuals to the effect that if the world had been otherwise he would have willed differently.[13]

But dependence cannot be construed in terms of counter-factuality alone. For counterfactuals can be contraposed. 'If I had bought those shares, I would be a rich man now' is equivalent to 'If I am not a rich man now, it is because I did not buy those shares': and the 'it is because' is witness to our sense that the counterfactual alone does not fully express dependency, which requires also some temporal element. 'If he had willed otherwise, things would have been different' is expressed not just by means of *if . . . then*, but by means of tenses, with the antecedent in the pluperfect, and so clearly earlier than the perfect conditional of the 'would have'. Dependence, in fact, is expressed not only by counterfactuals, but by time. What is later depends on what is earlier, and not *vice versa*. It is temporal order that distinguished the two limbs of Walker's

[12] Walker, *Kant*, p. 38.
[13] *Ibid.*, p. 40.

characterization of free agency, namely, 'If he had willed otherwise, things would have been different' and 'if the world had been otherwise he would have willed differently'. In a timeless world we could not distinguish the effect of the agent on the world and that of the world on the agent: the same counterfactual would express both that the agent was effective and that he was determined; or rather, we should have no concept of the will, and should construe the counterfactuals linking the subject with the world as perhaps expressing the way his perceptions depended on the world rather than the way his decisions shaped the world.

The argument from the maturation of probabilities is admittedly less strong than that from agency. There is no sense of purpose, no argument to distinguish one set of counterfactuals from their contrapositives. But still there is the argument from anisotropy. It is hard to believe that the parameter we have to posit to make sense of the collapse of probabilities could just as well be spatial as temporal. Space is many-dimensional and isotropic, whereas the transition from probability to actuality is directed and one-dimensional. If we are to have any collapse of probability into certainty it must be the same thing which at one stage had only a probability of being such and such and at another stage was certainly such and such or certainly not such and such: the parameter must be one that things can continue the same over, and with a definite direction. The parameter would appear to be one-dimensional and anisotropic, and such that entities remain the same over it. It certainly points towards time.

The same arguments hold for possibility. Possibility pre-supposes actuality. I cannot know what it is for something to be possibly the case unless I know what it would be for it to be actually the case. In our normal approach we tend to use probabilities of scientific phenomena, because they can be quantified, and possibilities of human choices because they usually cannot. But for our present purposes possibilities are in the same case as probabilities. Once again,[14] we can think of

---

[14] See above, ch. 4, § (iv), pp. 76–7, and ch. 5, § (ii), pp. 91–2.

them as generalised probabilities: if the whole range of probabilities is the closed interval [0,1], we can take 0 itself to be impossibility, 1 to be necessity, and all the other probabilities in the open interval (0,1) to be contingent possibilities. Under this interpretation, what we have argued for probabilities applies to possibilities just as well.

Suppose, then, we have a set of propositions and a set of modal standpoints, and from some of the modal standpoints some of the propositions are possible. The argument from maturation suggests that if some proposition is possible from some standpoint, then there must be another from which it is not merely possible, but actual. It is not quite the principle of plenitude put forward in the ancient world. Not everything possible is realised, but rather to say that something is possible is to commit oneself to its being under other conditions *either* true *or* false. There is a 'moment of truth' for each proposition at which it is either definitely true or else definitely false. That is to say, for every proposition p, there is a modal standpoint, i, such that $True_i(p) \vee Fls_i(p)$, or, what comes to the same thing, $True_i(p) \vee True_i(-p)$.[15] Although we are concerned with the modal standpoint from which the proposition p is either definitely true or definitely false, since we are not using indexed propositions, we cannot assume that $True_i(p)$ be always defined. It is therefore convenient to strengthen $True_i(p)$ to $L_i p$, which be always assigned a truth-value, and consider $L_i p \vee L_i -p$. We can conveniently abbreviate this as $U_i p$ (it is Unalterably the case that either p or not-p), and its negation $M_i p$ & $M_i -p$ as $O_i p$ (it is Open whether p or not-p).

The two operators U_i and O_i cannot be properly viewed as being themselves modal operators, since U_i does not satisfy κ (it may be unalterably true that if I contracted Green Monkey disease yesterday I am going to die tomorrow, and it is either unalterably true or else unalterably false that I contracted Green Monkey disease yesterday, but it does not follow that it is unalterably true that I shall die tomorrow or else unalterably

---

[15] See above, ch. 8., § (iii), definition (4), p. 148.

false – it may be that I have not contracted Green Monkey disease, but might still be run over tomorrow morning by a car). And O_i clearly does not satisfy the rule of necessitation, since theorems are not open, but definitely true. Nevertheless it is worth noticing that both the operators are transitive, in that U_iU_ip implies U_ip, and O_iO_ip implies O_ip; and also that U_i is quasi-T-like in that U_ip_i is always true.

The moment of truth does not automatically constitute an ordering of modal standpoints, but it goes some of the way. We can consider not only the moment of truth of an unmodalised proposition p, but of a modalised one, such as $L_ip \vee L_i{-}p$. We can consider it from some other modal standpoint, j, and ask whether from that standpoint it be on the one hand open or on the other unalterably true or unalterably false; that is whether $M_j(L_ip \vee L_i{-}p)$ & $M_j{-}(L_ip \vee L_i{-}p)$ or $L_j(L_ip \vee L_i{-}p) \vee L_j{-}(L_ip \vee L_i{-}p)$; that is, whether $O_j(O_ip)$ or $U_j(U_ip)$.[16] If the former, then $j < i$.[17] But the converse does not hold. This is because the moment of truth need not be contemporaneous with the event; in some modalities an event becomes inevitable or impossible well before its actual occurrence. I have missed the train while still waiting for a bus in a distant suburb. In the intervening period, while telephoning my apologies to my host, it is necessarily true that I have missed the train, L_ip , and hence that it is necessarily true or necessarily false that I have missed the train, $L_ip \vee L_i{-}p$, or in short, U_ip: but that was unalterably so from the time the bus failed to come; hence $L_j(L_ip \vee L_i{-}p) \vee L_j{-}(L_ip \vee L_i{-}p)$, and not $M_j(L_ip \vee L_i{-}p)$ & $M_j{-}(L_ip \vee L_i{-}p)$, or more succinctly, $U_j(U_ip)$ and not $O_j(U_ip)$, though clearly it is not the case that $i < j$.

We cannot establish a complete, linear ordering of modal standpoints granted only one modality and an exiguous stock of propositions. But there are many modalities, and many, very

---

[16] We should note that since O_ip is $-U_ip$, $U_j(U_ip)$ is the same as $U_j(O_ip)$ and $O_j(U_ip)$ is the same as $O_j(O_ip)$.

[17] It is easily seen that openness is transitive: O_ip implies M_ip, so $O_j(O_ip)$ implies M_jM_ip, which in turn implies M_jp; O_ip also implies $M_i{-}p$, so $O_j(O_ip)$ implies $M_jM_i{-}p$, which in turn implies $M_j{-}p$; hence O_jp

many, propositions. I have missed the train while still waiting for a bus in a distant suburb. But there could be a miracle: a magic carpet or a friend with a helicopter might turn up and whisk me to the station in a twinkling of an eye. Although so far as practical possibility goes I have missed the train, there are wider possibilities in which my being at the station in time is still a Q-accessible possible world. Of these different modalities, logical possibility is the widest. It provides an outer boundary for accessibility relations, and hence for temporal accessibility relations. Given any particular modality, a proposition is open up until its moment of truth, and unalterably true or unalterably false thereafter: so its moment of truth is the earliest moment, i, at which $U_i p$, that is, $L_i p \lor L_i{-}p$. We then consider all the successively wider modalities that satisfy 4 but not B, L, L′, L″, etc., with possibly different moments of truth, L_i, L'_j, L''_k, etc., where $i \le j \le k$ etc., and the upper limit of this sequence, l say, is contemporaneous with the event described by p, so that we can ascribe to p the index l, and write p_l. Similarly, granted just one modality, as between two different modal standpoints, we can consider not just whether or not $L_j(L_i p \lor L_i{-}p) \lor L_j{-}(L_i p \lor L_i{-}p)$, but whether or not $L_j(L_i p' \lor L_i{-}p') \lor L_j{-}(L_i p' \lor L_i{-}p')$, whether or not $L_j(L_i p'' \lor L_i{-}p'') \lor L_j{-}(L_i p'' \lor L_i{-}p'')$, etc. It is a reasonable assumption that, given the very large number of propositions we might entertain, we can always find for any two distinct modal standpoints some proposition which will have a different modal aspect. That is, quantifying over modal standpoints and over propositions,

$$(Ai)(Aj)(-(i=j) \to$$
$$(Vp)(((L_i p \lor L_i{-}p) \ \& \ -(L_j p \lor L_j{-}p)) \lor (-(L_i p \lor L_i{-}p) \ \& \ (L_j p \lor L_j{-}p)))$$

or, more simply,

$$(Ai)(Aj)(-(i=j) \to (Vp)((U_i p \ \& \ O_j p) \lor (U_j p \ \& \ O_i p)).$$

Granted this assumption, we can order the modal standpoints completely, that is, establish the law of trichotomy:

$$(Ai)(Aj)((i=j) \lor (i < j) \lor (i > j)),$$

which gives us a linear ordering of modal standpoints, and if we can have sufficiently wide possibilities, a criterion of contemporaneity between modal standpoints and particular propositions. We have thus justified our previous use of linearly ordered indices, s, t, u, v, to index temporal standpoints and chronologically definite propositions.[18]

(vii) The Topology of Time

That time should be cyclic is highly implausible. It runs counter to our ideas of personal identity that we could come round again to the self-same situation and make the self-same decision. Rather than admit cyclic time we will go to almost any lengths, even to that of an endless recurrence of events in serial time, to keep time as we think it ought to be. It is a merit of the modal approach that it rules out cyclic time absolutely. For in cyclic time the accessibility relation Q would be symmetric, and there would be no fundamental difference between past and future. Against this, however, Prior distinguishes global from local directedness, and cites Hamblin's 'east-west' tense-logic in which 'California is east but not west of Sydney, and west but not east of Manchester,'[19] to show that some local difference between past and future could be maintained even though time were cyclic. But the local difference between past and future extends to a global one. Q^{-1} is linear, and Q is not: if time were cyclic there would be one and only one Q^{-1}-accessible course of previous history coming back on itself, which would in this way distinguish one Q-accessible course of future history from all the others, thus denying the openness of the future and the real possibility of choosing one of the others rather than the one which would come round to where we were.

---

[18] See above, ch. 6, § (i), pp. 106–7.
[19] A. N. Prior, *Past, Present and Future*, Oxford, 1967, pp. 63–6. For a philosopher who takes cyclic time seriously, see W. H. Newton-Smith, *The Structure of Time*, London, 1980.

A more intuitive argument is that of newness. If time were cyclic it would be true that there is nothing new under the sun, since for any p, it would be true that, in Priorean terms, $p \rightarrow Pp$. One counter-instance would be enough to refute it: if we may quantify over propositions, it would be enough to have

$$(Vp)(p \ \& \ H{-}p).$$

And this we can assert: there is something new under the sun: I, being unique, am new; you, being unique, are new; so is everyone else. Every undergraduate believes that when he came up to the university, there was a fundamental change in the intellectual scene. Nor is that just the conceit of youth. It is part of our concept of an autonomous rational agent. We believe that human beings are autonomous and creative, and can fashion in their lives and in their thoughts something entirely fresh. And again this is not just the generalised conceit of the human race: the inexhaustible power of reason to develop new modes of ratiocination has been vindicated in this century by some surprising results in mathematical logic.[20]

In the terminology of modal standpoints it is natural to use the Open operator, O_i, and to claim that there is, not just sometimes but at all times, some opportunity open and not yet unalterably fixed. We can express the doctrine that there are always new opportunities by

$$(Ai)(Vp_j)O_ip_j.$$

This implies more than that time is not cyclic. It asserts that it is serial in the strong sense that it is unending. For every modal standpoint there is another further down the line, which is contemporaneous with some proposition, and is, therefore, in some modality its moment of truth; and so there is no last

---

[20] See above, ch. 4, § (i). For a brief, and relatively non-technical, account of Gödel's theorem, see Ernest Nagel and James R. Newman, *Gödel's Proof, New York, 1958, London, 1959*; for its use as an argument against determinism, see J. R. Lucas, *The Freedom of the Will*, Oxford, 1970.

modal standpoint. If every accessibility relation is in this strong sense serial, the corresponding modalities satisfy axiom D,[21]

D
$$\Box p \to \Diamond p.$$

It also follows from $O_i p_j$ that $-(\Diamond_i p_j \to \Box_i p_j)$, and hence that the thesis D*

D*
$$\Diamond p \to \Box p$$

does not hold for the future operators $\Box$ and $\Diamond$, that is, that future modalities are not degenerate.[22] We can also argue the other way. If future modalities are never degenerate, then D but not D* must hold, and there can never be an end of time. The parallel argument about the beginning of time, however, is not valid. This is because U, unlike O, is quasi-T-like, in the sense that

$U_i p_i$ is always true, whereas
$O_i p_i$ is always false.

If there were a beginning of time, then for any proposition about the state of affairs obtaining then, p_0,

$U_0 p_0$ would be true.

From the modal point of view, the present is like the past, both being actual. For the present, as for the past, (that which is necessarily is), when it is, as Aristotle held.[23] The present is the first moment of definiteness, which remains unalterable thereafter, and is, so to speak, the very last moment of the past, while the future is always not yet. Any requirement of actuality as characteristic of the past is satisfied by the present too, and so if there were a moment that had no past, the requirement of actuality would be satisfied then nonetheless. It is the fact that the future is modally live that leads us to the conclusion that there will always be a future.

[21] See above, ch. 5, § (iv), p. 97.
[22] See above, ch. 5, § (i), p. 89.
[23] See above, ch. 7, § (i), p. 124, n.5.

The future is open in a further, topological, sense. There is no last moment before which an open possibility of a particular proposition collapses into being definitely true or definitely false:[24]

$$(Ai)(Aj)(Vk)((k>i) \ \& \ (O_ip_j \rightarrow O_kp_j))$$

Until the moment of truth, when the die is cast definitely one way or the other, there is a perpetual possibility of averting what is feared, or losing what is sought: there is always the possibility of a slip between the cup and the lip.

It follows that time is dense, and not only dense but continuous. For, though we deny that there is a first moment of the future, we claim that there is always a last moment of the present-and-past, namely the present. Aristotle had discussed whether continuous change had a first or a last instant, and the Schoolmen wrote many treatises *De primo et ultimo instanti*. There are four tenable positions: that there is always a last moment of non-change and a first moment of change; that there is always a last moment of non-change, but not a first moment of change; that there is always a first moment of change, but not a last moment of non-change; and that there may be neither. The first of these positions holds that time is discrete, the last that it is dense but not continuous, and the other two that it is continuous. We have argued for the second, but we should note that there is an important difference between Aristotle's indefinite 'change' (or our 'future') and a definite particular future event, p_j. There is a last moment of the present-and-past, a last moment, that is *before* the future, but with any particular event in the future there will be some intervening period of the future before that particular event occurs.

The doctrine of the ever-present present, which would establish that there is always a last moment of the present-and-past, although intuitively attractive, needs support. It can be supported by two arguments, neither incontrovertible but together telling.

[24] *Physics*, VI, 5, 253^{b}32–236^{a}27 and VIII, 8, 263^{b}9–26.

The first employs a Dedekind cut. Given any two distinct standpoints, i and j, there is some proposition, p_l, that divides them, so that, for example, $O_i p_l$ but $U_j p_l$. Every standpoint, k, can be classified similarly, and will either be such that $O_k p_l$ or $U_k p_l$. We thus have a Dedekind cut of standpoints, and can identify the cut with the index, l, of the proposition p_l.

A second argument arises from a variant on Cochiarella's axiom[25]

$$\mathbf{GH}(\mathbf{H}p \to \mathbf{FH}p) \to (\mathbf{H}p \to \mathbf{GH}p)$$

If it is always going to be the case that it has always been the case that (if it has always been the case that p it will sometime be the case that it has always been the case that p) then (if it has always been the case that p it is always going to be the case that it always has been the case that p). Essentially, it embodies a principle of complete mathematical induction. There are two premisses, the first, $\mathbf{GH}(\mathbf{H}p \to \mathbf{FH}p)$, expressing the inductive step, the second, $\mathbf{H}p$ the basis of the induction, which together imply that p be always true, provided time is continuous, but would fail, at for instance $\sqrt{2}$, if time were merely dense. For in that case the two premisses might hold for for all times less than $\sqrt{2}$, but not for any times greater. Thus the axiom secures continuity. If we supposed it false, we should be committed to a very cautious principle of extrapolation without generalisation: it would be all right to maintain that something that had been green hitherto would continue a little longer always having been green but not that it was always going to be green. It would be a principle of if green yesterday, then green after yesterday – maybe today, maybe tomorrow – but perhaps blue in the fullness of time. Such a principle is not inconsistent. Many articles in learned journals have been exploring just that possibility. But it suffers from something like ω-inconsistency, and it would be irrational to be willing always to extrapolate and yet not to generalise. We do not think there are break-points in the world around us. And in ruling out that

---

[25] Prior, *Past, Present and Future*, p. 72.

possibility we are committed to some principle of the togetherness of nature, which in the case of time amounts to a requirement of continuity.

(viii) Possible Worlds *versus* Actual Time

Prior talked of branching time, and this seemed counter-intuitive. We feel that possible courses of events may branch, but that time must be linear. Next year I may marry Jane and I may marry Jean, and these two possibilities constitute quite different life histories for me, but there is only one next year. Time, like space, we naturally think, is a matrix, within which different events may occur, but not in itself different.

But it is difficult to give an adequate account of time, and positivist scruples make some philosophers queasy about ascribing properties to time apart from the events that occur within it. There was a continual ambiguity in Tree Semantics whether temporal standpoints were were to be identified with possible worlds, w_{it} *etc.*, or the dates that indexed a whole set of possible worlds. The principle of maturation applies to propositions, shared by a whole set of possible worlds. We do not hold that every possible world must sooner or later be actualised, as the ancient principle of plenitude would suggest, but that every proposition must sooner or later come true or come false. Each proposition is true of half of the possible worlds at the time, its negation being true of the other half, and the set of all propositions that come true at a particular date characterizes just one possible world, which alone is actualised at the date in question, all others fading away as might-have-beens. This was why it was correct to speak of *the* Tree of t, rather than the Tree of w_{it}, which was only *a* Tree of t, one of many each with its node at a particular possible world, w_{it}, w_{jt}, w_{kt}, etc. With sub-trees, however, the case was different. They were not yet actual Trees, and could not be characterized by date alone, but needed also some specification of the possible state of affairs supposed to be obtaining then. Their hold on actuality was thus more tenuous: only in as much as they were sub-trees of the actual Tree of t could they be granted firm ontological status or confidently talked about. It was not enough that particular theses held of all the sub-trees of the Tree of t that had their node at u; it was incumbent also to argue that these together constituted the Tree of t, so that what was true for all the u sub-trees of the Tree of t could properly be said to be true of the Tree of t itself.

A similar intimation of actuality is shown in our urge to use past tenses for counter-factual hypothetical propositions. We find it hard to stomach counter-factual possible worlds, and attempt to do so by back-tracking to a temporal standpoint which was, and therefore still is, actual, and from which the possible world envisaged really was possible. Such a move is not always legitimate. As we noted, some counterfactuals, for example those involving a breach of the laws of nature, never were temporally possible.[26] But the urge to adopt this particular usage remains strong, and indicates a feeling for the high ontological status of actual time as opposed to the shadowy existence of counter-factual possible worlds.

It is actuality that distinguished Reichenbach's S, the time of utterance, from his R, the reference point referring to whatever temporal standpoint we choose to take up. Our minds can wander freely and we can envisage many possibilities, but when the moment of truth comes, only some of what we had envisaged will turn out to be true in the event, and our situation is what it is, and not necessarily as we had supposed. It is from the mandatory actual standpoint that we have to speak, and it is from there that we phrase our discourse. Though we are free to view events from any temporal standpoint we please, and can construct appropriate RE propositions to communicate to others, in actually expressing them we have to view them from the standpoint we are obliged to occupy. The RE propositions are nested in a non-optional outer S shell.

[26] See above, ch. 8, § (vi), p. 157.

We can talk diachronically, that is across time, remembering or reading at one time what had been said or written at another. The common core of what is said, or denied, or argued about, is the RE proposition, but the way it is expressed at different times varies with the perspective in which we are obliged, at the time of utterance, to view it. So it is that if we wish to gainsay yesterday's utterance 'There will be a sea battle tomorrow' in the future simple tense (S–R,E), we say now that 'There isn't one' in the present tense (S,R,E), and later on 'There wasn't one' in the aorist (R,E–S).

The mandatory shift of temporal standpoint gives us a sense not only of the passage of time, but of its ineluctability. We cannot choose what time it is now, and hence we cannot choose the outer perspective we have to have on everything we talk about. The necessity seems to impinge all the more sharply because it contrasts with the freedom we have to take up any temporal standpoint we please within the outer shell, and view events either contemporaneously, or from before, or from after, the time they take place. Whereas our minds are free to roam over all ages, our bodies are imprisoned, as we tell ourselves, within a prison of temporality, so that our tongues are compelled to speak, if they speak at all, only at the time we find ourselves to be living and speaking at. Looked at like this, it is a prison. But if we consider the concomitants of agency, we see that we could be agents only if we could choose, and exclude some possibilities in actualising others, and could achieve something only if the past, as well as the present, were necessary and unalterable, and that acting at a particular time is a corollary of being a particular agent, not just a unit but an individual, with a mind of his own to make up, and a contribution to the course of events that, too, is all his own.

11

The Vulnerability of God

(i) Theology

We return to metaphysics and theology. Our understanding of time has deep implications for our view of reality and God, the ultimate reality. If time is a perpetual becoming, a weaving rather than an unrolling, we cannot take a static view of the universe, but must see it as dynamic, in which something is always happening, vague possibilities crystallizing out into sharp actuality. Reality is through and through temporal. Equally, God is temporal, though not merely that. Although we can properly say that God is more than merely temporal, that He transcends time, and to that extent is beyond and outside time, we cannot say that He is timeless, or that for Him there is no difference between future and past.

A long tradition of Christian theology maintains the opposite, that God is timeless and changeless. To think otherwise, it is suggested, is to relapse into crude anthropomorphism, and to read the New Testament in the light of the Old. Boethius, having seen that Origen's reconciliation of Divine fore-knowledge with human freedom would not work,[1] resorted to the timelessness of God as the only way of preventing an insoluble problem being posed.[2] Only if God is distanced from

---

[1] Ch. 3, § (i), pp. 31–4.

[2] *De Consolatione Philosophiae*, V, vi. There is a hint of this solution in St Augustine;'s response (*Ad Simplicianum*, II, Quaest. ii) to the difficulty felt by Simplician; over the text in I Samuel 15:4, 'It repenteth me that I have set up

the world of space and time can He be acquitted of responsibility for the terrible things that happen within it – an omnipotent Deity who operated in time could have intervened to prevent the holocaust, and if He did not, showed Himself thereby to be insensitive to the sufferings of others, and less than perfectly benevolent.

On the face of it, such a claim runs entirely counter to the biblical account of God, which reveals Him as acting in history, speaking by the prophets, answering prayer, and concerned with everything that happens in the world. Admittedly, it is an anthropomorphic conception of God. But why not? God made man in His own image, and was incarnate as a man. Although we need to be careful, in thinking about God, not to worship images that are merely the creatures of our own fashioning, it cannot be argued that being in the form of a man is contrary to the divine nature. Nor is the New Testament understanding of God a non-interventionist one. God intervened in the course of events in sending His Son to live among us, in speaking at His baptism, in His transfiguration, and above all in raising Him from the dead. He intervenes less crudely than He is sometimes portrayed as doing in the Old Testament, but He intervenes none the less. The problem of evil remains a problem however we understand omnipotence. There may be reasons why a good God should not intervene, even to prevent the holocaust, but they must be moral reasons, not metaphysical reasons, if they are to be adequate.[3]

There are further, more philosophical, arguments from the nature of God, His omniscience, omnipotence and perfection, which again seem to require that He be not compromised by any taint of temporality or transitoriness. If we reject their conclusion, which we must if we are to be faithful to the biblical

---

Saul to be king'; he denies that God can have foreknowledge, because foreknowledge is knowledge of the future, and God transcends – 'supergreditur' – all times.

[3] See below, § (vii).

revelation of God and our own understanding of human agency, we must profoundly rethink our understanding of the nature of God.

(ii) Timelessness

The timelessness of God is often confused with the changelessness of God, but it is an entirely different doctrine. If God is timeless, it makes no sense to ask whether He might change, just as it makes no sense to ask whether e might cease to be an irrational number: if God is changeless, it makes perfectly good sense to ask the question, and the answer will always be No, just as it makes perfect sense to ask whether pigs can fly, even though the answer is uniformly negative.

The doctrine that God is timeless stems from the Greek conception of God. Plato at one stage in his life thought that the ultimate reality was the form of the good, which, like the other forms, was a timeless abstract entity, very much like those of mathematics, to be described impersonally as θεῖον [*theion*], divine, rather than ὁ Θεός [*ho Theos*] God. Although in his later life, Plato did come to speak of God, the tradition of an impersonal absolute continued in Platonist and Neo-platonist thinking, and was absorbed into mediaeval theology, so much so that the Schoolmen talked of the ultimate reality as *ens realissimUM* in the neuter, rather than *Ens Realissimus*.

The timelessness of God can be argued for on two grounds. Boethius draws an analogy with space. We do not say that God is in space, but rather that all space is present to God. *Eodem modo*, in the same way, we should not say that God is in time, but that all time is present to God.[4] But time is not like space, and being temporally present is not like being spatially present. In its temporal application the word 'present' can govern either some word denoting an instant or some word denoting an interval. In the latter sense it is true that the whole of time is,

[4] *De Trinitate*, IV, ll. 59–77.

from God's point of view, present time,[5] but that presupposes that God is not timeless but temporal. It is the former sense that leads Boethius to take over Plotinus' definition of eternity as a static, frozen, block view of time, which God could view from outside, but in which, were He within it, He would be be in some way limited and confined.

Time is not like space, but still the argument from limitation carries weight. To say that the ultimate reality is within time seems to impute a restriction incompatible with ultimacy. Far better, it would seem, to say that it is outside, or better still, beyond, time. Cosmologists attempting to account for the origin of the universe have sometimes toyed with the idea of a non-temporal entity generating time and space at the beginning of time, and St Augustine held that time was created by God at the creation of the world. Such a view is tenable, but denies that the ultimate reality is a person. To be a person is to be conscious and to be an agent. Time is the concomitant of consciousness and the condition of agency. If the argument from limitation be valid, there can be no God, but only some more tenuous entity, not to be conceived of in personal terms, not to be conceived of in any terms whatever. For to characterize is to limit. In applying one predicate rather than another, we are thereby delimiting the character of what we are talking about, and in saying that it has some features, we indicate thereby that it lacks others.

If, therefore, we refuse on principle to allow any limitation whatsoever on the nature of God, we are reduced to saying nothing about Him. The Greek Fathers started along the path of characterizing God only negatively, and in the apophatic theology of the Cappadocians came near to the reverent agnosticism of the Neoplatonists who simply spoke of τὸ ἐκεῖ [*to ekei*], the beyond, without saying anything about it. Absolute infinitude, taken to mean utter unboundedness, leads to total indefiniteness. But though we should be very

---

[5] See below, § (iv), and J. R. Lucas, *A Treatise on Time and Space*, London, 1973, §§3,4,55, pp. 17–25, 300–7.

cautious in saying anything about the ultimate reality, knowing how inadequate our concepts are to such a task, we cannot say nothing. That would be a counsel of despair. We want to know, and need to make up our minds in order to live our lives on the basis of what really is the case rather than what is not. We must, therefore, try to formulate, albeit only tentatively, our best conclusions about the fundamental nature of reality, acknowledging, with St Augustine, that we only say what we say because to say nothing would be more misleading still.[6]

If we are to characterize God at all. we must say that He is personal, and if personal then temporal, and if temporal then in some sense in time, not outside it. That God cannot alter the past should be seen not as a lamentable lack of power on the part of the Almighty, but as a corollary of His being an agent. To be an agent is to be crystallizing potentiality into actuality, thereby making it unalterable thereafter. No unalterability, no agency. To bewail unalterability is like bewailing God's inability to make a world so big that He cannot move it, and to project an inconsistency in our concepts into an inability in what they are being applied to.

Traditionally time, like space, has been thought of as a thing, and therefore something created by God, and not existing before He created it. Before God created time, we are sometimes encouraged to think, time did not exist, and then God certainly was a timeless being. But the very formulation of the statement that purports to tell us what God was like before time existed is incoherent. Time is not a thing that God might or might not create, but a category, a necessary concomitant of the existence of a personal being, though not of a mathematical entity. This is not to say that time is an independent category, existing independently of God. It exists because of God: not because of some act of will on His part, but because of His nature: if the ultimate reality is personal, then it follows that time must exist. God did not make time, but time stems from God.

---

[6] St Augustine, *De Trinitate*, V.9.10.

(iii) Changelessness

The Fathers cited a number of biblical texts to prove that God was changeless. Psalm 102, vv. 26 and 27, reads:

They shall perish, but thou shalt endure: they shall all wax old as doth a garment. And as a vesture shalt thou change them, and they shall be changed: but thou art the same, and thy years shall not fail.

Malachi 3 v.6 declares

For I am the Lord. I change not; therefore ye sons of Jacob are not consumed.

And St James 1 v.17 tells us

Every good gift and every perfect gift is from above, and cometh down from the Father of lights, with whom is no variableness, neither shadow of turning.

But these texts do not establish the case. For one thing the texts themselves do not naturally bear the meaning that God is absolutely changeless, but only that He does not change in certain, highly relevant respects. The Psalmist is saying in Psalm 102, vv.23–28, as in Psalm 90, v.2, that whereas man is mortal, God is immortal, and Malachi is saying that God is not fickle, and will not abandon the children of Israel, even though they have often abandoned Him. In the same spirit St James goes on to tell his readers how God did something, and imbued us with the word of truth. Moreover, the texts cited are very few. They can hardly outweigh the whole thrust of the biblical record, which is an account of God both caring and knowing about the world, even the five sparrows, which at one time had not yet been, and later had been, sold for two farthings, and intervening in the world, doing things, saying things, hearing prayers, and sometimes changing His mind.[7]

[7] See esp., Oscar Cullman, *Christ and Time*, tr. F. W. Filson, London, 1962; see also James Barr, *Biblical Words for Time*, London, 1969, esp. ch. 3.

The Fathers, aware of the natural reading of the Bible, sought to interpret it away in the light of their philosophical commitments. St Thomas Aquinas follows St Augustine and St Anselm in construing the mental life of God entirely behaviouristically.

> When certain human passions are predicated of the Godhead metaphorically, this is done because of a likeness in the effect. Hence a thing that is with us a sign of some passion is signified metaphorically in God under the name of passion. Thus with us it is usual for an angry man to punish, so that punishment becomes an expression of anger. Therefore punishment itself is signified with anger, when anger is attributed to God.[8]

It is surprising to find Christian theologians taking up the same position as Ryle did in his *The Concept of Mind*, and it is open to the same objections. A purely behaviourist account of the actions and feelings of another makes them unintelligible. We cannot understand why the different manifestations of conceit that Ryle brilliantly portrays[9] should be all manifestations of the same character *trait* unless we have the idea that the conceited person is always thinking about himself. Equally we cannot understand the actions and feelings attributed to God in the Bible unless we can, at least to a limited extent, have some idea of God reaching a decision and caring about what happens. And once those are allowed, temporality creeps in, and the biblical texts can no longer be interpreted as revealing a changeless God. The changelessness of God is not to be naturally read out of the Bible, but rather was read into it in the light of certain philosophical assumptions about the nature of God.

---

[8] *Summa Theologiae*, I.Q.19, art.11, resp.; quoted Nicholas Wolterstorff;, 'Suffering Love', in Thomas V. Morris, *God, the Good, and Christian Life*, Notre Dame Press, 1988, p. 235, n. 22; Wolterstorff gives careful consideration to the arguments of the Fathers and the Schoolmen, but reaches conclusions very similar to those of this chapter. See also Anselm;, *Proslogion*, 8, and J. K. Mozley;, *The Impassibility of God*, Cambridge, 1926; and A. J. Moen, *God, Time and the Limits of Omniscience*, Oxford D.Phil. thesis, 1979, ch. 3.

[9] Gilbert Ryle, *The Concept of Mind*, London, 1949, p. 171.

Plato had argued that God must be changeless. For if God changed, the change must be either for the better or the worse: if for the better, then God had previously been less good than He might have been, and so was not the most perfect possible being and if for the worse, then He was subsequently less good than He might have been, and so was not the most perfect possible being. The argument is an argument from perfection, and, like the arguments from God's infinity adduced in the previous section, needs to be examined with care. Although at first sight attractive, it depends on the assumption that there is one linear scale of excellence, so that any two different states of affairs can be compared and ordered, one better, one worse. But this is not so. There are many different excellences, and it is perfectly possible to change from one sort of excellence, that is the best of its kind, to another, which is the best of its, quite different, kind.

Other arguments for changelessness are based on the doctrine of impassibility and the inappropriateness of God changing His mind. It is one thing for God to change, but quite another for Him to change at the behest of external circumstance and, in particular, the behest of mere man. Not only for the Greeks, but even for the Jews, it seems a denial of the majesty of God, 'for God is not as a man, that he should be threatened, neither as the son of man, that he should be turned by intreaty'.[10] These arguments, however, stem not from the very concept of God, as the ultimate reality, but from differing views about what He is actually like. There is a divergence of understanding between the Greek and the Judaeo-Christian understanding of the nature of God, and to this we shall need to return later.[11]

(iv) Eternity

Eternity is often thought of as being a very long time, or alternatively, absolutely timeless. But those concepts we already

[10] Judith 8 : 16b.
[11] See below, §§ (vi) and (viii).

possess, and there is no need to annex the word 'eternity' to either of them. The reason we need to talk of eternity is because we want to approach time from the Godward, rather than the human, side. Once again, great caution is required. We cannot say what God's experience of time is like. But just as it is useful in other fields of philosophical enquiry to consider the God's-eye view of things, so it is illuminating to consider the formal requirements of a God's-eye view of time.

There are difficulties in ascribing a particular date to an omniscient being. We are tempted to say that everything He knows is immediately present to His consciousness, and therefore is contemporaneous with His awareness of it, and so is simultaneous with everything else He knows. But this cannot be true. Although when we are talking about intervals, we can say that for God the present interval extends over the whole of time, so that there is no time that is for Him in the remote past or the remote future, when we are talking of the instantaneous present, we must say that for Him, as for us, only the present instant is present. Else He cannot hear our prayers when we pray, or answer them in due season. If God is, as Christians believe, able to communicate with human beings, for whom different times at different times are present, then a time that once was but now is not present to a prayerful petitioner, once was but now is not present to God. If God heeded St Augustine's prayers, He heard them and responded to them in St Augustine's lifetime, centuries after He spoke with Moses in the burning bush, and centuries before He was moved by the prayers of John Wesley. The claim that all three events *be* present, in the instantaneous sense, together in the mind of God leads to the conclusion that they are all simultaneous with one another.[12] Only one instant is present to God, and all the rest are either past or future, just the same as with men. God's knowledge of the present and past is immediate and fresh, but not, in the case of the past, contemporaneous.

[12] Anthony Kenny, 'Divine Foreknowledge and Human Freedom', in Anthony Kenny, ed., *Aquinas: a Collection of Critical Essays*, London, 1969, p. 264.

The Special Theory of Relativity has been thought to offer a way of escaping this conclusion. It shows us that the concept of simultaneity is not an absolute one, but a frame-dependent one: that is, in asking whether two events are simultaneous with each other or not, we need to specify with respect to what frame of reference we are asking the question, and it is perfectly possible for two events to be simultaneous with respect of one, and not with respect to another, frame of reference. In one frame of reference, it is thought, St Augustine's praying might be simultaneous with God's speaking with Moses in the burning bush, and in another with the praying of John Wesley. But no escape is possible along this route. For one thing, Moses, St Augustine and John Wesley all inhabited pretty well the same frame of reference and had substantially the same canon of simultaneity. And for another, it is only in ascribing dates to very distant events that a change of frame of reference makes a big difference to the date to be ascribed, and we are considering prayers uttered and answered within comparatively restricted periods of time and regions of space. Wherever the earth was located in the solar system, God's speaking with Moses in the burning bush was in the Absolute Past of St Augustine's praying, which in turn was in the Absolute Past of John Wesley's praying, and no choice of frame of reference could make it seem otherwise.[13]

It has also been suggested that some special sort of simultaneity, 'ET-simultaneity', is appropriate for discussing the relation of divine and human events.[14] But ET-simultaneity is not transitive,[15] and therefore not an equivalence relation, and so cannot be regarded as a genuine species of simultaneity. It is essentially denying the temporality of God, and hence His being a person at all.

---

[13] For the problems raised by the Special Theory of Relativity, see more fully, J. R. Lucas and P. E. Hodgson, *Spacetime and Electromagnetism*, Oxford, 1990, §§ 2.7, 3.1, 3.9.

[14] Eleonore Stump and Norman Kretzmann, 'Eternity', *The Journal of Philosophy*, 78, 1981, pp. 429–58; reprinted in Thomas V. Morris, ed., *The Concept of God*, Oxford, 1987, pp. 219–52.

[15] Ibid., p. 440/231.

There is a further, much more profound, difficulty arising from the frame-dependence of simultaneity. Different inertial frames of reference have different canons of simultaneity, so that events that are simultaneous in one frame of reference will not be simultaneous in another: and, according to the equivalence principle, each of these different frames of reference is just as good as any other. But if God is omniscient, He must know the occurrence of events at the time they occur, and must be able to tell whether two widely separated events are simultaneous with each other, or whether one of them is before the other. His omniscience must embody some canon of simultaneity, and must canonize it as being *the* canon, contrary to the fundamental equivalence principle of relativity.

Einstein's equivalence principle arises from the Lorentz transformation, which is the transformation we need to adopt if we are to describe electromagnetic phenomena so as to bring out those features that are essentially the same from whatever point of view they are described. It is the right principle to adopt when we are considering electromagnetic interactions, and perhaps also for weak, strong, and gravitational interactions. But this is no warrant for making the equivalence principle an absolute one, or extending it beyond the bounds of physics. Although we need to say that so far as electromagnetism is concerned, all inertial frames are on a par, we should not be flying in the face of empirical facts if we were led by theological or metaphysical considerations or by considerations from some other part of physics to prefer one frame above all others.

It is illuminating to compare Einstein's equivalence principle with Galileo's. Galileo's equivalence principle lays down that, so far as Newtonian mechanics is concerned, all inertial frames are on a par. Although they all have the same canon of simultaneity, each determines a different state of rest. So far as absolute rest is concerned, Newtonian mechanics and the Special Theory of Relativity are exactly the same: we cannot tell, in the one case by any mechanical test, in the other by any any electromagnetic observation, whether an inertial frame is at rest or not. But this does not mean that the concept of absolute rest is meaningless, nor that a frame at absolute rest could not be picked out by some other means. Newton believed, on metaphysical and theological grounds, that there was one, and was inclined to think that the centre of mass of the solar system was in a state of absolute rest. Much later the Michelson–Morley experiment was undertaken in order to determine the earth's motion through the electromagnetic aether. If it had yielded a positive result, it would have enabled us to determine a frame of reference in which the aether was at rest, and it would have been natural to pick out that frame as being absolutely at rest. Though Galileo's equivalence principle would still have held good for mechanical phenomena, there would have been none the less a frame of reference that was preferred when a wider range of phenomena were under consideration.

The divine canon of simultaneity implicit in the instantaneous acquisition of knowledge by an omniscient being is not incompatible with the Special Theory of Relativity, but does lead to there being a divinely preferred frame of reference in the same way as the aether might have constituted an electromagnetically preferred one. In each case there would be a preferred frame of reference, which would be compatible, in the one case with the Newtonian and in the other with the electromagnetic, laws of physics. It would not make any difference within Newtonian mechanics or electromagnetic theory, but would be of significance from a more general point of view.

An omniscient being can know all there is to know of temporal reality. He has immediate, undimmed knowledge of the present and past. He can view things from any temporal standpoint He pleases, experiencing both what is occurring and what has occurred, and envisaging the sort of thing that would occur or might have occurred, were certain conditions fulfilled. He has what the Schoolmen termed Middle Knowledge, though only in general terms, since the actual decisions of particular individuals are not there to be known, unless and until they are actually taken.

We can thus give a coherent account of God's temporal relations with His creation. It is a matter of distinguishing what is conceptually necessary, if there is to be a coherent concept at all, from what is only a contingent limitation, due to human weakness and imperfection. God is not limited in the way we are. We forget: God does not. We are impatient: God is not. We fail to think ahead: God does not. Although temporal predicates can be applied to God in a full sense, and though God changes in some respects with the passage of time, He does not grow old and wear out, and instead of being subject to time, as we are, can properly be said to transcend it.

(iv) Foreknowledge and Forebelief

Boethius was led to the conclusion that God was timeless as an escape from the problem of foreknowledge and free will.[16] In fact, however, foreknowledge, rightly understood, is compatible with freedom, partly because it is first-personal, partly because it invokes a number of different modalities, not all of them Procrustean, and is therefore inherently defeasible.

I know what I am going to do tomorrow. You may too. God also can. Each of us is privy to his own counsels, and often by his explicit avowals makes others privy too. If God is about my path, and about my bed, and if there is not a word in my tongue He does not altogether know,[17] then He will have a fair idea of my future course of action, far from complete, of course, since I have not made up my mind about many things, and do not know myself what I am going to do, but enough none the less to predict some things with a fair degree of certainty. Even if God did not know the secrets of men's hearts, but only what they explicitly told Him or implied in their importunate petitions, He would still be better informed than most of us, who nonetheless manage to predict quite a number of future events with success.

[16] See above, § (i), p. 209 and n. 2.
[17] Psalms 139: 2,3.

But this sort of foreknowledge does not foreclose freedom. It is either entirely, or else vicariously, first-personal foreknowledge, resting on an internal rather than an external modality, which, rather than restricting one's freedom of action, gives it definition and structure. I can decide what I am going to do but that does not mean that thereafter I am absolutely not free not to do it, but only with respect to my upholding my previous decision. Only an internal necessity binds me. I cannot do other than do it, except at the cost of abandoning my own previous decision, which I am rationally reluctant to do.

But of course I always can. My decisions, however binding, are not irreversible. Internal internal modalities are always subject to a *homine volente* escape clause. And then my own and other men's well-founded predictions will be falsified, and what had seemed to be foreknowledge will prove to have been merely mistaken, though well-grounded, belief. Foreknowledge can Cambridgely become, *ex post facto*, not knowledge at all, and for this reason lacks the Procrustean power to bind the future.

God, having vicarious first-personal knowledge of our future actions, and knowing also the ways and waywardness of man, can foreknow much, but in a non-threatening, because in a not external and not infallible, way, and always subject, so far as particular actions of particular individuals are concerned, to a *homine volente* clause. But still we may press specific questions. Did God foreknow the Second World War? The misery of the Germans during the slump, the wickedness of Hitler, the blindness and irresolution of Britain and France, were grounds enough for predicting war, if not over the Sudetenland then over Danzig, if not over Danzig then over the Saar, or Denmark, or Alsace-Lorraine. But always there was the possibility of things going differently. If Britain had not caved in at Munich, and if the General Staff had then succeeded in removing Hitler, then the inevitable would not have happened, and foreknowledge, even God's foreknowledge, would have proved not knowledge after all.

Foreknowledge may be massaged away by suitable modalities and Cambridge change, but some unease remains. Whatever account we give of knowledge, belief is not defeasible. It is a hard, unalterable fact that you believed I would go to Professor Strawson's lecture tomorrow, and if, when tomorrow comes, I do not go, then there is no question of your not having really believed it: you did believe it, and were wrong. So too with God. If God had a belief yesterday that I should do something, and I do not do it, then God was wrong in His belief. If God had believed that Peter would deny Him, and Peter, when questioned, rose courageously to the occasion, and confessed that he was one of Jesus' disciples, then God would have been wrong. Nor is this a merely hypothetical possibility. In the Old Testament He is often represented as changing His mind, and not carrying out the threatened consequences of the Jews' bad behaviour He had announced that He would bring about. Elijah foretold evil for Ahab in the name of the Lord. 'I will bring evil upon thee, and will take away thy posterity, and will cut off every male of the house of Ahab, bond or free, in Israel.' . . . but when he heard those words, he rent his clothes, and put sackcloth upon his flesh, and fasted and lay in the sackcloth, and went about dejectedly. And the word of the Lord came to Elijah saying 'Seest thou how Ahab has humbleth himself before me? Because he humbleth himself before me, I will not bring the evil in his days, but in his son's days will I bring evil upon his house.'[18]

The natural reading is that God changed his mind. He did not speak mendaciously to Elijah the first time, but at that time intended to sweep away the house of Ahab in his own lifetime so that he could appreciate his punishment, but later, in the light of his penitence, suspended part of the sentence for the remainder of his life. And he might have done more, had he repented more. Suppose Ahab had not merely fasted and put on sackcloth himself, but, besides giving back the vineyard to Naboth's heirs, had universalised the prescription, 'Let neither man nor beast, herd nor flock, taste anything; let them not feed, nor drink water, but let man and beast be covered with sack-

[18] I Kings 21: 17–24, 27–9.

cloth, and let them cry mightily to God;' . . . thinking to himself 'who can tell if God will turn and repent, and turn away from his fierce anger, so that we perish not?'[19] Certainly, when Hezekiah had been told by Isaiah that he was going to die, God heard his prayer and changed His mind, giving Hezekiah another fifteen years of life.[20]

Such changes of heart were a problem for thoughtful theists. The author of the book of Jonah points out that an unrelenting, God who secures His absolute veracity at the cost of never showing mercy, would be less perfect than one whose ears are open to our prayers.[21] In the New Testament the good news is that if we repent, we can escape the consequences of our ill-doing that would otherwise ensue. 'Except ye repent, ye shall all likewise perish'.[22] The clear picture is of a God who can change His mind, and is prepared to make prophecies, issued in His name and on His explicit commands, come false.

Most thinkers have thought it utterly incompatible with God's perfection that He should ever be mistaken – *quod nefas judico*, says Boethius.[23] Not only Boethius, but Origen and St Augustine before him,[24] and except for Socinus almost all Christian thinkers since, have thought that it would derogate from God's greatness if He could be mistaken. Much argument is needed if we are to reject that conclusion, and show that God can be not only fallible but actually mistaken without its derogating from His excellence and perfection.

[19] cf. Jonah 3: 7–9.
[20] 2 Kings 20: 1–6.
[21] Psalms 34: 15.
[22] St Luke 13: 3 and 5.
[23] *De Consolatione Philosophiae*, V, iii, ll. 6–16.
[24] Origen, *apud* Eusebius, *Praeparatio Evangelica*, bk. 6, c. 11, p. 286, in J-P. Migne, *Patrologia Graeca*, Paris, 1837, 21, 492; St Augustine, *De Libero Arbitrio*, III, ii, 4, quoted above, ch. 3, section (vi), p. 48; cf. *City of God*, V, ch. ix.

(vi) Perfection

God is perfect, but perfection, like infinity, is easily misunderstood. As we saw earlier with infallibility,[25] it is easy, by insisting on one apparent excellence, to narrow, rather than widen, the range of divine competence. So, more generally, as Findlay, Kenny and Blumenfeld have pointed out,[26] it is easy to set one superlative against another, and conclude that no entity at all could possibly be superlative in all respects. If God is absolutely omnipotent, and can always change His mind, He cannot be absolutely omniscient, and know infallibly that He never will. In any case, He cannot make a world so big He cannot move it, nor know absolutely everything, false propositions as well as true.

We need to construe the the *omni* of omnipotence and omniscience, not in terms of some inconsistent, absolute all, but negatively, as contrasting with various forms of non-omnipotence and non-omniscience. I am not omnipotent because there are lots of things I cannot do, and it is a defect of mine that I cannot. I cannot write Greek elegiacs, I cannot jump over the moon, I cannot swim across the Behring Strait. But other people can, and it is only because I am not as good as they that I cannot. Equally, I am not omniscient: I do not know French, I do not know Twister theory, I do not know American history. Other people do, and I should be less ignorant if I did too. In other cases, however, it is no defect of mine that I cannot do, or do not know, something. I cannot make a flag that is red and green all over, not on account of any inadequacy of mine, but because it is logically impossible task. It is no skin off my nose that I cannot alter the past – it is not something I ought to

[25] Ch. 3, § (vi), pp. 48–9.
[26] N. J. Findlay, 'Can God's Existence Be Disproved?', *Mind*, 57, 1948, pp. 108–18; Anthony Kenny, *The God of the Philosophers*, Oxford, 1979; David Blumenfeld, 'On the Compossibility of the Divine Attributes', *Philosophical Studies*, 34, 1978, pp. 91–103, reprinted in Morris, ed., *The Concept of God*, pp. 201–15.

remedy, or could try to overcome with effort or careful attention to the teaching of Professor Dummett. Where the question 'Why can't you?' or 'Why don't you know?' has to be met with a confession of inability or ignorance on my part, it is reasonable to look for God's not being subject to any such limitation. But where it is not due to an incapacity of the person but is in the nature of the case that something cannot be done or known, then it is no derogation from God that He cannot do it or know it either.[27] God cannot sin: God cannot know first-hand what it is to have sinned: God cannot infallibly know what He is going to do until He has made up His mind – else omniscience would have foreclosed His freedom, and curtailed His omnipotence – and God cannot, so long as He has created us free and autonomous agents, infallibly know what we are going to do until we have done it. But this is no imperfection, but a corollary of His creative love.

The theological superlative is a potent source of error. There is a sense in which God is the mostest. There is a sense in which God is the bestest. God is the ultimate reality. But there are many sorts of greatness, many sorts of goodness, incommensurate and not always compatible, and in ascribing maximality to God, we need to have in mind in what way He is the greatest or the best. Questions of ontology and value are involved, and as we come to understand more fully what it really is to exist and what values are truly worth espousing, we advance also in our understanding of the perfection of God.

---

[27] Compare Thomas P. Flint and Alfred J. Freddoso, 'Maximal Power', in Alfred J. Freddoso, ed., *Existence and Nature of God*, reprinted in Morris, ed., *The Concept of God*, p. 151: 'Therefore . . .there will be some state of affairs . . . which even an omnipotent agent is incapable of actualizing. And since this inability results solely from the logically necessary truth that one being cannot causally determine how another will freely act, it should not be viewed . . . as a kind of inability which disqualifies an agent from ranking as omnipotent.'

(vii) Providence

An omnipotent God could have chosen to create a world in which not everything could be known, even by Him. Einstein said he could not believe in a dice-playing God, but unless quantum mechanics is logically inconsistent, an almighty God could have chosen to make a world that exemplified the laws of quantum mechanics. And if He did, then He would be unable to have detailed knowledge of its future development.

Many thinkers have felt that it would run counter to the doctrine of providence if the details of the future course of events were not foreknown and fore-ordained. But such a doctrine, though widely held, is un-Christian. Apart from leaving no room for human freedom, it poses the problem of evil in irresoluble form, and subverts the moral teaching of Jesus.

There are many tribulations in human life that are not attributable to the wickedness of men. Earthquakes, famines, disease and death are evils. Good may come out of them in the fortitude with which they are met, or the sympathy they evoke, but they are evils none the less. If God 'freely and knowingly plans, orders and provides for all the effects that constitute His artefact, the created universe with its entire history, and executes His chosen plan by playing an active causal role sufficient to ensure its exact realisation',[28] then He is directly and immediately responsible for these evils. And then the project of justifying the ways of God to man becomes impossible to carry through.

Many thinkers have set their face against any such project. There is no searching of God's understanding, and He giveth no account of any of His matters. We cannot question the inscrutable decrees of God's providence, and should simply

---

[28] *Luis de Molina, On Divine Foreknowledge: Part IV of the Concordia,* translated, with an introduction and notes, by Alfred J. Freddoso, Ithaca, 1988, p. 3.

trust Him that, despite any appearances to the contrary, all will be well. Such a response, though expressing an important insight, does not capture the main thrust of the Christian revelation. Admittedly, it is important to emphasize the great gap between our very limited understanding and the profundity of the Divine Reason itself. There is a tendency for modern man, and especially modern Angry Young Man, to tell God what he is, and is not, prepared to accept, and to call God to account for Himself to his complete satisfaction. He needs to be answered out of the whirlwind. But he is not typical of the whole human race, and to insist upon the nothingness of man in comparison with the greatness of God belies the concern for His children expressed by God in the Old Covenant, and His sending His Son to save them in the New. Although there is a great gap between selfish, unredeemed humanity and the perfection of God, there is not intended to be a complete barrier between man and God – God became man in order that man might become like God – and the breaking down of the barrier opens the way to, and indeed is partly constituted by, the asking of questions. The faithful Christian, caring about the immense suffering many men have to undergo, must wonder why a good, all-powerful God should allow such terrible things to happen.

It could be that God is improving us or using us. Theages was saved by ill health from dissipation. A touch of arthritis may make us more sympathetic to the weakness of others. Only if I am ill can doctors have the opportunity to make me better and nurses the occasion to care for me. Only at my funeral will old friends renew old acquaintanceships. But such a justification, however applicable in general, cannot cover every particular case. Not every disaster can be accounted for as a providential means to a greater good. I am not invariably improved by disease or death. Although sometimes good may come out of ill, the loss remains and is real. Nor is it consonant with God's fatherly care for His children simply to use them as means towards some external end.

It could be that God is punishing us. That is what the Jews thought. But Jesus taught otherwise. Neither the cripple nor his parents had done wrong; neither the casualties of the collapse of the tower of Siloam, nor the victims of Herod's vendetta against Judaism, were sinners more than the rest of us.[29] It is, furthermore, contrary to the thrust of the New Covenant to see everything that happens as God's immediate response to our previous actions. The rigid moralism of the Pharisees could accommodate the Old Understanding that if we walk in God's ways it will be well for us, but such a legalism destroys the spontaneity of the heart and all personal relations. Unless God sends the rain on the unjust as well as the just, the disinterested love of justice cannot be made manifest or distinguished from meteorological prudence. We need to live in a world in which we do not always get what we deserve, so that we can exercise the good will uncontaminated by ulterior motives. I cannot be friends on a strict *quid pro quo* basis. If whenever I do a good turn, it earns me a good turn in return, my good will is merged in the self-interested motive of doing well by number one. I can do business on that basis, but not manifest friendship. In order to be friends there needs to be some separation between the nicely calculated less and more of merit and reward on the one hand and the free expression of feelings and attitudes on the other. To enter into personal relations I need room to be myself, not constrained by prudential reckonings with providence. It follows that the Christian view of providence must be, on purely theological grounds, different from the traditional one.

Not everything that happens can be attributed directly to the detailed decision of God. Although He knows how many hairs I have on my head, He has not decided how many there shall be.[30] He distances Himself from the detailed control of the course of events in order, among other things, to give us the freedom of manoeuvre we need both to be moral agents and to go beyond morality into the realm of personal relations.

[29] St John 9 : 2, St Luke 13 : 1–5.
[30] St Luke 12 : 7. The parallel passage in St Matthew 10 : 30 is, however, more predestinarian in tone. That the point at issue is one of God's concern rather than control is one of many I owe to D. J. Bartholemew, *God of Chance*, London, 1984.

Although He could, and perhaps on occasion does, intervene directly to avert a particular evil or guide things towards an appropriate end, He could not do it often or as a matter of course without destroying the conditions of freedom in which alone man could come to, and exercise, mature autonomy.

Questions of theodicy and providence remain. If we deny the doctrine of detailed providence, then we can attribute suffering to accident rather than to God, but still ask why God created such an accident-prone world, and what reason we have to think that He cares for us at all. The general considerations outlined earlier are a partial answer to the first question: accident-prone-ness is a concomitant of freedom: only through trial and error can we develop as self-reliant and autonomous individuals: only in an accident-prone world can we manifest sympathy and fortitude. There still remains a justified complaint on the part of latterday Jobs, but the terms of the complaint are altered, and the possibility of a further answer is open.

God's ordering of the world does not have to be detailed in order to manifest His goodness towards us. God provides for us also in securing generally beneficent conditions for our existence, and in there being certain tendencies in natural phenomena and human affairs that work to our advantage, often redressing some of the ill consequences of our bad decisions. God's providence is shown not in the fact that everything happens as it does, but in some good things happening when they might well have not happened. The providential ordering of the world is shown both in its general arrangements being conducive to our welfare, and in that setbacks, though real, can characteristically be overcome. In an earlier work I suggested that instead of thinking of God's providence as a sort of blue-print, with the inevitably Procrustean overtones of that metaphor, we should liken it to the Persian rug-maker, who lets his children work at one end while he does the other. The children make mistakes, but so great is his skill that he adapts the design at his end to take into account each error at the children's end, and works it into a new, constantly up-dated pattern.[31] The analogy is helpful so far as the relatively rare direct interventions of God are concerned, and the, perhaps more frequent, occasions when men are guided by God to do His will. But we need also to note a certain self-correcting tendency in the course of events, whereby if one thing fails another is likely to happen that will bring about the same effect. Were it not for this, the universe might get out of hand and become beyond saving. But the element of chaos introduced by chance and men's arbitrary bad decisions is not the only factor at work in a universe inhabited by men who are also partly rational and sometimes well-intentioned. There is a natural propensity for those governed by self-will to lose contact with reality, and ultimately to compass their own destruction, as Hitler, Amin and Galtieri did, and for those who are sensitive to the will of God to be able to work together in achieving their reasonable and realistic ends. The good we do is often taken up and amplified by other men: our evil plans are often self-frustrating, and our selfish failures interred with our bones. We may thwart God's purposes for a season, but in the long run the pervasive pressures of rationality and love will circumvent our petty resistances and secure a wider measure of cooperation than our self-isolating selfishness can defeat.

(viii) Vulnerability and Value

The perfection of God raises questions of value. Christian values go beyond those of ordinary morality. Love, rather than duty, is the keynote, and love is many-faceted. Instead of the monolithic scale of values assumed by Plato and most moralists, there are many different forms of fulfilment, and many different goals we should set ourselves to achieve with the passing of

---

[31] I have tried to work out the analogy more fully in J. R. Lucas, *Freedom and Grace*, London, 1976, chs. 4 and 5, esp. p. 39. See also Jacques Maritain, *God and the Permission of Evil*, cited by Flint and Freddoso, 'Maximal Power', in Freddoso, ed., *Existence and Nature of God*, reprinted in Morris, ed., *The Concept of God*, p. 163, n. 30.

time. We are moral agents, but not mere units of morality, whose sole job is to carry out the behests of the moral law, as in the dreary moral determinism preached by the Stoics and by Kant. Though bounded by moral considerations, fulfilment for the Christian is personal fulfilment, different for different individuals.

If individuals are valuable in their own right, as children are in the eyes of their father, the traditional doctrine of impassibility loses its appeal. It was thought to be incompatible with the ultimacy of God that He should be moved by anything outside Himself. But if God has created independent centres of value, He is not being constrained by some external force, if He then values them for what they are and what they become, independently of any further choice of His. Having created men in His own image, He has endowed their choices and predilections with a value in their own right. A father does not merely leave his children free to make up their own minds for themselves on occasion: he so much respects their choices that the fact that they want something is valuable in his eyes too. Likewise God is guided in the things He wants to happen by the things we want to happen. But this is not the constraint of external necessity, but the free first-personal choice of creative love.

Love is not only creative, but vulnerable. If I care for somebody I can be hurt. God, on the Christian view, is highly passible and was hurt. Instead of the impassive Buddha untroubled by the tribulations of mortal existence, the Christians see God on a cross: instead of the Aristotelian ideal of a self-sufficient God who devotes His time to enjoying the contemplation of His own excellence, the Christians worship a God who shared the human condition and came among us

The greatness of God is understood very differently on this view from the traditional account of His absolute perfection. God is a father, first and foremost, and His kingdom, power and glory are that of a father rather than an absolute monarch. Power, as ordinarily understood, is not the great good we think it is, but must, as Gregory of Nyssa pointed out, be understood in an altered perspective.

That the omnipotence of God's divine nature should have had strength to descend to the lowliness of humanity furnishes a more manifest proof of power than even the supernatural character of the miracles. . . . It is not the vastness of the heavens and the bright shining of the constellations, the order of the universe and the unbroken administration over all existence, that so manifestly displays the transcendent power of Deity as the condescension to the weakness of our nature in the way in which the sublimity is seen in lowliness, and yet the loftiness descends not.[32]

And much as it can be an exercise of omnipotence for God to limit Himself and make Himself vulnerable to the will of others, so the value that regards knowledge as a good can be more fully realised by forgoing the possibility of being a complete know-all, and creating a world in which the future actions of others can often only be surmised, and sometimes not even that. If God created man in His own image, He must have created him capable of new initiatives and new insights which cannot be precisely or infallibly foreknown, but which give to the future a perpetual freshness as the inexhaustible variety of possible thoughts and actions, on the part of His children as well as Himself, crystallizes into actuality.

[32] *Oratio Catechetica*, in J-P. Migne, *Patrologia Graeca-Latina*, 24; quoted A. M. Ramsey, 'Christian Belief – an Underlying Essence', *Religious Studies*, 11, 1975, p. 198.

Index